Fundamentals of EM Design of Radar Absorbing Structures (RAS)

Hema Singh • Ebison Duraisingh Daniel J
Harish Singh Rawat • Reshma George

Fundamentals of EM Design of Radar Absorbing Structures (RAS)

 Springer

Hema Singh
Centre for Electromagnetics (CEM)
CSIR-National Aerospace Laboratories
Bangalore, Karnataka, India

Ebison Duraisingh Daniel J
Centre for Electromagnetics (CEM)
CSIR-National Aerospace Laboratories
Bangalore, Karnataka, India

Harish Singh Rawat
Centre for Electromagnetics (CEM)
CSIR-National Aerospace Laboratories
Bangalore, Karnataka, India

Reshma George
Centre for Electromagnetics (CEM)
CSIR-National Aerospace Laboratories
Bangalore, Karnataka, India

ISBN 978-981-10-5079-4   ISBN 978-981-10-5080-0   (eBook)
DOI 10.1007/978-981-10-5080-0

Library of Congress Control Number: 2017945085

Printed on acid-free paper

This Springer imprint is published by Springer Nature
The registered company is Springer Nature Singapore Pte Ltd.
The registered company address is: 152 Beach Road, #21-01/04 Gateway East, Singapore 189721, Singapore

To
Late Dr. R.M. Jha

Abstract

Specifically the EM absorbers are in demand for the frequency range 1–40 GHz, because of their requirement in electromagnetic interference (EMI) shielding and countermeasure to radar detection. In defence sector, stealth technology demands such EM absorbers towards radar cross section (RCS) reduction. Total absorption of impinging electromagnetic (EM) wave requires the suppression of all propagation phenomena, viz., reflection, transmission, and scattering. The absorber can be a dielectric material or dielectric combined with metal plates placed at different intervals. The EM absorption is not difficult to achieve by increasing the volume of the absorber material and shaping its geometry, provided there is no limitation of space and weight. However practically, the space on airborne platforms is so limited that the designer has to consider various design factors, such as weight, thickness, absorptivity, environmental resistance, mechanical strength, etc. This brief presents a detailed analytical formulation, step-by-step design procedure for EM design of radar absorbing structures (RAS). Both the equivalent circuit model and Smith chart approach are discussed with illustrations. It provides the reader a clear understanding of the steps involved in designing multilayered RAS as per the desired specifications. This book will be helpful for beginners, academicians, and R&D engineers working in the field of RAS design and development.

Keywords Radar absorbing structure • EM design • Absorption • Equivalent circuit • Smith chart • Reflection coefficient • Admittance

Preface

In the present era, low observability is one of the critical requirements in aerospace sector, especially related to defense. The stealth technology essentially relates to shaping and usage of radar absorbing materials (RAM) or radar absorbing structures (RAS). The performance of such radar cross section (RCS) reduction techniques is limited by the bandwidth constraints, payload requirements, and other structural issues. Moreover, with advancement of materials science, the structure geometry no longer remains key decisive factor toward stealth. There is a major contribution of RAM and RAS in reducing the radar signatures. These RAM/RAS not only contribute in reducing RCS of vehicles such as aircraft or ship but also serves the purpose of reducing or even eliminating the presence of spurious radiations in the environment, caused by electronic equipment operating at high frequencies. EM absorbers are specially designed materials which can inhibit transmission or reflection of incident wave. The absorber can be a dielectric material or dielectric combined with metal plates placed at different intervals. The EM absorption is not difficult to achieve by increasing the volume of the absorber material and shaping its geometry, provided there is no limitation of space and weight. However, practically, the space on airborne platforms is so limited that the designer has to consider various design factors, such as weight, thickness, absorptivity, environmental resistance, mechanical strength, etc. This document presents a detailed analytical formulation and design procedure for EM design of radar absorbing structures (RAS). Both the equivalent circuit model and Smith chart approach are discussed. This work provides the reader a clear understanding of the steps involved in designing multilayered RAS as per the desired specifications.

Bangalore, Karnataka, India   Hema Singh

Acknowledgments

We would like to thank Mr. Jitendra J. Jadhav, Director, CSIR-National Aerospace Laboratories, Bangalore, for the permission to write this SpringerBrief.

We would also like to acknowledge valuable suggestions from our colleagues at the Centre for Electromagnetics, especially, Dr. R.U. Nair and Mr. K.S. Venu, and their invaluable support during the course of writing this book. We would like to thank the CEM project staff, Ms. Phalguni Mathur, for assisting in editing the manuscript.

If not for the concerted support and encouragement of Swati Mehershi, Executive Editor, Applied Sciences & Engineering, of Springer, it would not have been possible to bring out this book within such a short span of time. We very much appreciate the continued support of Ms. Aparajita Singh at Springer toward bringing out this brief.

About the Book

Specifically the EM absorbers are in demand for the frequency range of 1–20 GHz, because of their requirement in electromagnetic interference (EMI) shielding and countermeasure to radar detection. Further the EMI/EMC is one of the important issues in industries related to high-speed wireless data communication systems, wireless local area networks (LANs), mobile and satellite communication systems, all-time money (ATM) machines, etc., operating in microwave frequency range. In defense sector, stealth technology demands such EM absorbers toward radar cross section (RCS) reduction.

Total absorption of impinging electromagnetic (EM) wave requires the suppression of all propagation phenomena, viz., reflection, transmission, and scattering. The absorber can be a dielectric material or dielectric combined with metal plates placed at different intervals. This document presents a detailed analytical formulation and step-by-step design procedure for EM design of radar absorbing structures (RAS). Both the equivalent circuit model and Smith chart approach are discussed with illustrations. This book provides the reader a clear understanding of the steps involved in designing multilayered RAS as per the desired specifications. This book will be helpful for beginners, academicians, and R&D engineers working in the field of RAS design and development.

Contents

List of Figures

List of Tables

Chapter 1
Introduction

The main purpose of absorbers is to inhibit the reflections of incident electromagnetic (EM) wave. The rapid developments in the field of communication and radar systems directed the attention of researchers toward EM absorbers. During initial stages, absorbing structures were mainly used to avoid electromagnetic interference problem. Since low observability became a critical requirement in aerospace sector, the application of absorbers got broadened. This led to the development of radar absorbing material (RAM) or radar absorbing structure (RAS). The classic absorbers like Salisbury screen, proposed long back, consist of resistive sheet placed a quarter wavelength apart from the ground plane. It is a good example for an absorber with simple structure and design procedure. However, the Salisbury screen has a drawback of bandwidth constraints. Jaumann absorber, considered as a higher version of Salisbury screen, consists of several resistive sheets placed quarter wavelength apart from each other. Its performance compromises on total thickness, but provides broader bandwidth. In order to achieve wider bandwidth with reduced thickness, frequency selective surface (FSS) is included in the design of absorbers. A FSS consisting of metallic structures and dielectric layer exhibits resistive as well as reactive properties. This results in impedance matching over a wide bandwidth. In other words, broadband absorption can be achieved with thinner absorbers.

The frequency response of such absorbers is usually evaluated through circuit simulating software or by using numerical-based methods, viz., method of moments (MoM). These numerical methods provide less information related to the design procedure and are computationally expensive. In the design phase of product development, it is necessary to have fast yet reasonably accurate means of evaluating the performance. For simple structures like Jaumann and Salisbury screen, the transmission line model (TLM) of the absorber and Smith chart provide sufficient information about the frequency response and overall design. In order to get an immediate knowledge of electromagnetic properties of complex FSS structures, an equivalent circuit analysis can be adopted. Unlike full-wave simulation using

H. Singh et al., *Fundamentals of EM Design of Radar Absorbing Structures (RAS)*, DOI 10.1007/978-981-10-5080-0_1

circuit simulating software, TLM provides good insight into the physical and design properties of the periodic structure.

The aim of the present work is to provide basic understanding to the beginners about the absorber and its design procedures. This document has three sections. In Section 1, a simple Jaumann absorber is considered, and the design rules for the same till N layer is discussed. Section 2 explains how to build a simple model of absorber using an equivalent circuit model toward achieving basic design that meets the desired specifications. In Section 3, the design rules for absorber design using the Smith chart approach is explained. Once the basic design is done, one can go for full-wave simulation and more accurate modeling.

Chapter 2
Design of Radar Absorbing Structure

As mentioned earlier, EM absorbers are specially designed materials which can inhibit transmission or reflection of an incident wave. The absorber can be a dielectric material or dielectric combined with metal plates placed at different intervals. One knows that absorption happens when the effective impedance is matched to the free space impedance. In other words, in case of impedance matching, the wave incident on the absorber passes through the material without any reflection and gets absorbed.

Absorbers can be constructed by placing one or more resistive sheets in a stratified medium (Munk et al. 2007). The most effective design method is to place a homogenous resistive sheet at a distance of $\lambda/4$ in front of a perfectly conducting ground plane. Such an absorber is called as Salisbury screen and possesses a narrow absorption bandwidth (around 26%) (Kazemzadeh and Karlsson 2009). The bandwidth can be increased by adding resistive sheets to the medium, spaced $\lambda/4$ apart resulting in so called Jaumann absorber. The bandwidth can be enlarged further by placing suitable dielectric slabs between the resistive sheets particularly in front of the outermost resistive sheet. This absorber broadens the absorption bandwidth by compromising the overall thickness (Lee and Lee 2012). In other words a Jaumann absorber is a multilayered version of the Salisbury screen. The main components of a Jaumann absorber include (a) spacers, (b) resistive sheets, and (c) a metal plane, as shown in Fig. 2.1. A spacer along with the resistive sheet adjacent to its incident side is considered as a Jaumann section. Each of these components is infinite and homogenous in transverse dimensions, whereas spacers alone are homogenous in the axial direction (Toit 1994).

(a) **Spacers** The spacers are assumed to be identical lossless dielectrics, i.e., $\mu = \mu_0$, and completely specified by the dielectric constant ε_r, where $\varepsilon = \varepsilon_o \varepsilon_r$ with $\varepsilon_r \geq 1$.

The characteristic impedance of a spacer is expressed as

© The Author(s) 2018
H. Singh et al., *Fundamentals of EM Design of Radar Absorbing Structures (RAS)*,
DOI 10.1007/978-981-10-5080-0_2

Fig. 2.1 Topology of a
Jaumann absorber

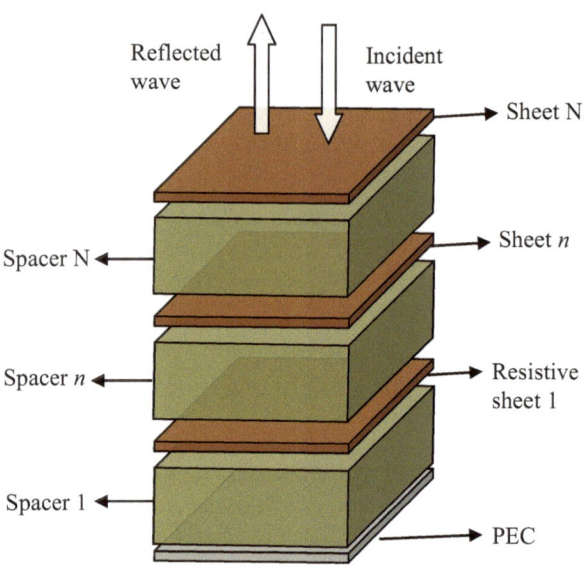

$$z_s = \frac{\eta_0}{\sqrt{\varepsilon_r}} \approx \frac{120\pi}{\sqrt{\varepsilon_r}}\,\Omega, \qquad (2.1)$$

It is noted that the characteristic impedance of dielectric spacer is equal or less than the intrinsic impedance of free space, η_0 (Toit 1994).

(b) Resistive Sheets Resistive sheets are assumed to have zero thickness and are completely specified by surface resistivity $R_{s,n}$ in Ω/sq. and lies in the range $0 < R_{s,n} < \infty$ where n=1, 2, …N.

(c) Metal Plane Metal plane acts as substrate on which dielectric and resistive sheets are placed. Here the metal plane includes a target object (whose reflection coefficient is to be reduced) and an infinitesimally thin perfect electric conductor (PEC).

Figure 2.2 presents a sequence of steps required for designing a RAS with arbitrary configuration.

2.1 Single-Layer Jaumann Absorber

A single-layer absorber commonly known as the Salisbury screen consists of a single resistive sheet, a dielectric layer, and a metal sheet, as shown in Fig. 2.3.

The corresponding dimensions used in design of a single-layer absorber are given in Table 2.1. The dielectric spacer in between resistive sheet and ground plane is taken as foam with permittivity value equal to that of air. The dielectric is

Fig. 2.2 Flowchart for designing RAS of arbitrary configuration

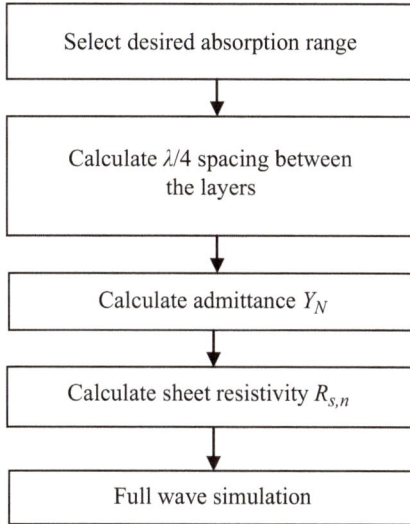

Fig. 2.3 Schematic of a single-layer absorber

Parameters	Layer 1
Dimension of patch (mm × mm)	6 × 6
Resistivity of square patch (Ω/Sq)	377
Width of dielectric (mm)	6
Thickness of dielectric (mm)	7.5

Table 2.1 Dimensions of a single-layer absorber

placed at a distance of d_1, and it is taken as $\lambda/4$ since the electric field is maximum at this point.

The front and transparent view of design of a single-layer absorber are shown in Fig. 2.4. The reflection coefficient of Salisbury screen is shown in Fig. 2.5.

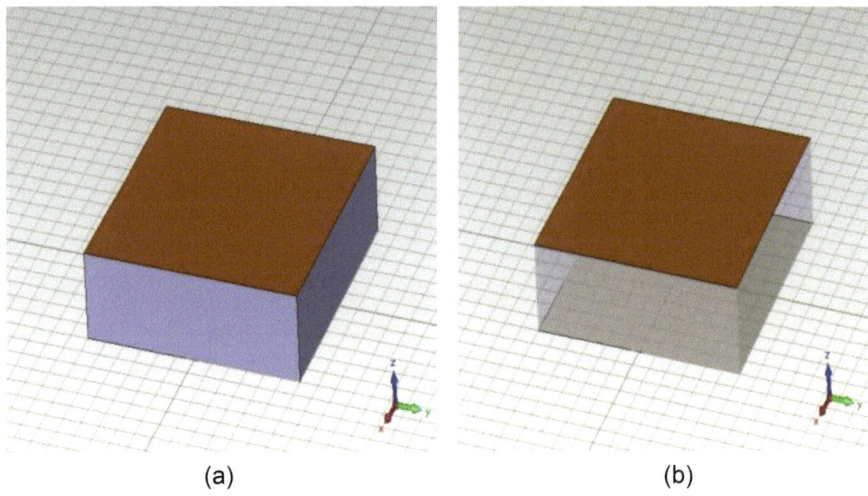

Fig. 2.4 Single-layer EM absorber, (**a**) front view, (**b**) transparent view

It is apparent that the reflection level obtained is below -20 dB around the center frequency, i.e., from 8.5 GHz to 11.5 GHz.

ABCD Matrix Approach For the sake of calculations, analysis, and better understanding, an absorber can be modeled as a transmission line circuit. The transmission line model for a single-layer Jaumann absorber is shown in Fig. 2.6. Such a section can be solved easily using cascade parameters of the voltage ABCD matrix T_i [Appendix A].

The input impedance and voltage reflection coefficient equations are derived using S as a frequency element in complex plane s, where $S = \tanh s$ and $s = j\omega$. The ABCD matrix for the above transmission line section is given by (Toit 1993)

$$\begin{bmatrix} V_1 \\ I_1 \end{bmatrix} = \begin{bmatrix} 1 & Z_c S \\ Y_n & Z_c Y_n S + 1 \end{bmatrix} \begin{bmatrix} V_2 \\ -I_2 \end{bmatrix} \tag{2.2}$$

This fundamental section of transmission line model for a single layer is more than enough to solve for Jaumann absorber with N sections. It is obtained by multiplying individual matrices of different sections arriving at a 2×2 matrix X_n:

$$X_N = \begin{bmatrix} A_N & B_N \\ C_N & D_N \end{bmatrix} = \prod_{n=0}^{N-1} T_{N-n} \tag{2.3}$$

Here,A_N, B_N, C_N, D_N are the matrix elements for N layers.
The input impedance of the structure is given by

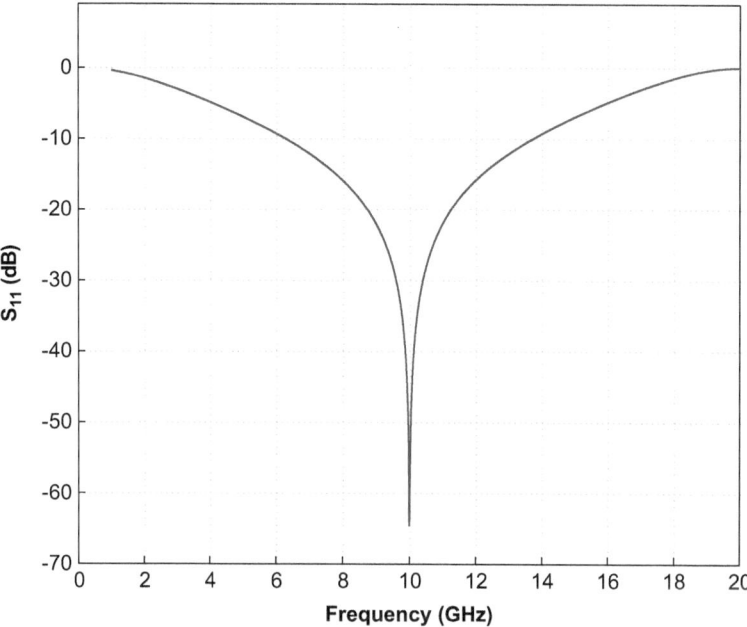

Fig. 2.5   Reflection coefficient of a single-layer absorber

Fig. 2.6   Isolated Jaumann section

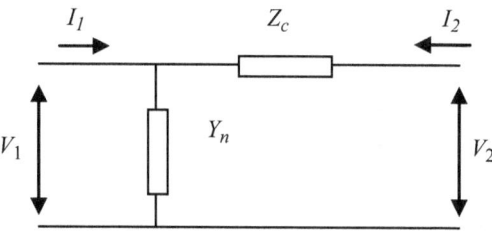

$$Z_{in} = \left.\frac{V_1}{I_1}\right|_{V_2=0} = \frac{B_N}{D_N}, \qquad (2.4)$$

and the reflection coefficient $\Gamma(S)$ is expressed as

$$\Gamma(S) = \frac{Z_{in} - 1}{Z_{in} + 1} = \frac{B_N - D_N}{B_N + D_N} \qquad (2.5)$$

The measureable reflection coefficient in terms of normalized physical frequency, ω, can be calculated by using (Toit 1993)

$$|\Gamma|^2 = \Gamma(S)\Gamma(-S)|_{S=j\tan\omega} \tag{2.6}$$

Z_{in} and Γ are rational functions of two real polynomials in S of order N. The coefficients of S is the sum of products of the admittance Y and the characteristic impedance Z_0. The details of derivation are given in the next section.

2.2 Double-Layer Jaumann Absorber

The schematic diagram of a double-layer Jaumann absorber is shown in Fig. 2.7. Each layer of absorber is represented by a matrix. Here matrix A and matrix B correspond to Layer 1 and Layer 2:

$$[A] = \begin{bmatrix} 1 & Z_cS \\ Y_1 & Z_cSY_1 + 1 \end{bmatrix} \quad \text{for Layer 1,} \tag{2.7}$$

$$[B] = \begin{bmatrix} 1 & Z_cS \\ Y_2 & Z_cSY_2 + 1 \end{bmatrix} \quad \text{for Layer 2,} \tag{2.8}$$

Using matrix multiplication $B \times A = \begin{bmatrix} A_N & B_N \\ C_N & D_N \end{bmatrix}$, the matrix elements B_N, D_N of ABCD matrix (for $N = 2$) are obtained as

$$B_2 = 2Z_cS + Z_c{}^2S^2Y_1 \tag{2.9}$$

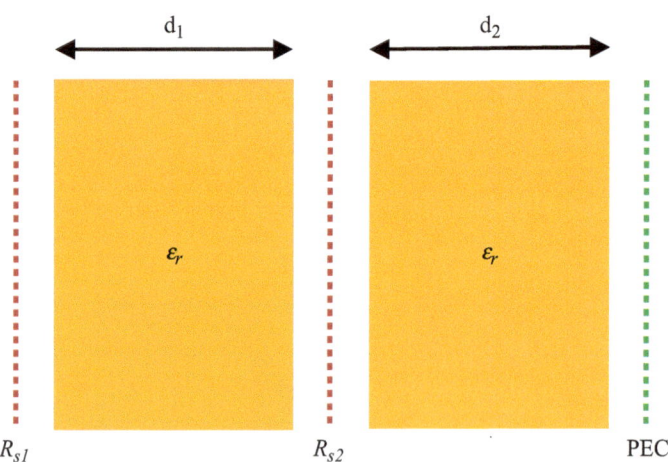

Fig. 2.7 Schematic diagram of a double-layer absorber

$$D_2 = Z_c^2 S^2 Y_1 Y_2 + Z_c S(Y_1 + Y_2) + 1 \tag{2.10}$$

Then the input impedance is obtained as

$$Z_{in} = \frac{B_N}{D_N} = \frac{B_2}{D_2} \tag{2.11}$$

The corresponding reflection coefficient for double layer is given by

$$\Gamma = \frac{B_N - D_N}{B_N +} {}^N D_N = \frac{B_2 - D_2}{B_2 + D_2} \tag{2.12}$$

Substituting the values of B_2 and D_2 in (Eq. 2.12), an equation in the general form is obtained:

$$\Gamma = \frac{a_1 S^2 + a_2 S - 1}{b_1 S^2 + b_2 S + 1} \tag{2.13}$$

where

$$a_1 = Z_c^2 Y_1 (1 - Y_2) \quad a_2 = Z_c (2 - Y_1 - 2Y_2) \tag{2.13a}$$
$$b_1 = Z_c^2 Y_1 (1 + Y_2) \quad b_2 = Z_c (2 + Y_1 + 2Y_2) \tag{2.13b}$$

A typical design of a two-layer Jaumann absorber is shown in Fig. 2.8.

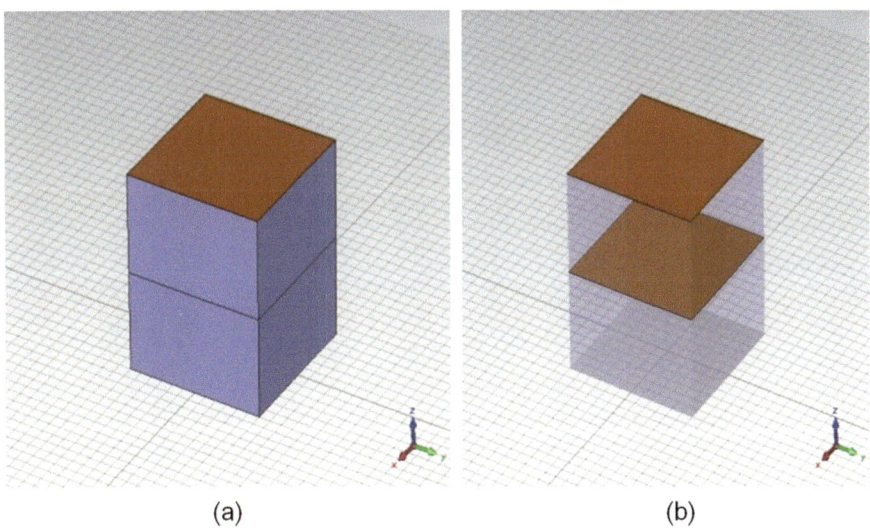

(a)   (b)

Fig. 2.8 Design of a two-layer absorber, (**a**) front view, (**b**) transparent view

A two-layer Jaumann absorber is designed by placing one more resistive layer on the top of a single-layer absorber (Fig. 2.4) at a distance of $\lambda/4$ from the first resistive sheet. In this design the electrical thickness remains the same since the operating frequency is not changed.

Calculation of Admittances Y_1 and Y_2 The resistive sheets above the dielectric spacer should be designed in such a way that effective impedance is matched to the free space impedance. Y_1 and Y_2 are obtained from (Eq. 2.13a), as explained in Appendix B. Then these admittances are used to find the effective impedance of corresponding layers.

From (Eq. 2.13) it is clear that there exist two solutions for Y_1 and Y_2:

$$\text{Solution A}: \quad \left\{ \begin{aligned} Y_1 &= \frac{-a_2 + \sqrt{\Delta}}{2Z_c} \\ Y_2 &= 1 + \frac{-a_2 - \sqrt{\Delta}}{4Z_c} \end{aligned} \right\} \tag{2.14}$$

$$\text{Solution B}: \quad \left\{ \begin{aligned} Y_1 &= \frac{-a_2 - \sqrt{\Delta}}{2Z_c} \\ Y_2 &= 1 + \frac{-a_2 + \sqrt{\Delta}}{4Z_c} \end{aligned} \right\} \text{where } \Delta = a_2{}^2 + 8a_1 \tag{2.15}$$

With $a_1 = 1/S^2$ and $a_2 = 0$, solution B always gives negative Y_1 that is not acceptable. Therefore, solution A is considered (Appendix B).

On simplifying (Eq. 2.14),

$$Y_1 = \frac{\sqrt{2}}{Z_c S}, \quad \text{and} \tag{2.16}$$

$$Y_2 = 1 - \frac{1}{\sqrt{2}Z_c S} \tag{2.17}$$

$$Z = Z_c S = \frac{\sqrt{\mu_r}}{\sqrt{\varepsilon_r}} \quad \text{where}, \mu_r, \varepsilon_r = 1, \ Z = 1 \ \Omega \text{ (for air)}$$

Thus, $Y_1 = \frac{\sqrt{2}}{1} = \sqrt{2} \ /\Omega$ and $Y_2 = 1 - \frac{1}{\sqrt{2}} = 0.2928/ \ \Omega$
This gives

$$R_{s,1} = \frac{377}{Y_1} = \frac{377}{1.414} = 266.619 \Omega/sq \quad R_{s,2} = \frac{377}{Y_2} = \frac{377}{0.2928} = 1287.56 \Omega/sq.$$

The parameters of a two-layer absorber are tabulated in Table 2.2. The relative permittivity of the dielectric spacer is taken as that of air. The thickness of the dielectric is $\lambda/4$ at center frequency of 10 GHz.

The reflection coefficient of a double-layer absorber (whose resistivity values are calculated from above derivation) is shown in Fig. 2.9. It is evident that a two-layer

Table 2.2 Dimensions of a double-layer absorber

Parameters	Layer 1	Layer 2
Dimension of the patch (mm × mm)	6 × 6	6 × 6
Resistivity of square patch (Ω/Sq)	266.619	1287.56
Width of dielectric (mm)	6	6
Thickness of dielectric (mm)	7.5	7.5

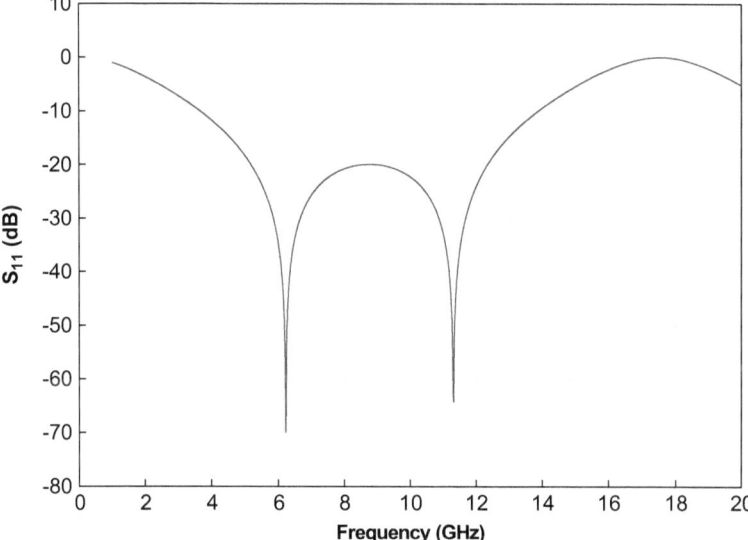

Fig. 2.9 Reflection coefficient of a double-layer absorber

absorber is better than a single-layer absorber in terms of absorption ($A = 1\text{-}R\text{-}T$) as well as bandwidth. However the trade-off is in terms of thickness and payload.

2.3 Design Procedure of *N*-Layer Absorber

From previous section, one may notice that the numerator of reflection coefficient is sufficient to find the admittance value (as per design objective, $\Gamma = 0$). When the number of layers is increased, the complexity in calculations for admittance for each layer also increases. Therefore, in order to compensate this problem, a general equation to find the "n" admittance values for N layers is required.

It is known that reflection coefficient $\Gamma(S) = \dfrac{Z_{in} - 1}{Z_{in} + 1} = \dfrac{B_N - D_N}{B_N + D_N}$ $\qquad$ (2.18)

Writing a generalized form of an equation for the reflection coefficient, one gets

$$\Gamma(S) = \frac{\sum_{n=1}^{N} a_n S^n - 1}{\sum_{n=1}^{N} b_n S^n + 1} \tag{2.18a}$$

with a_n and b_n positive real finite coefficients.

For design procedure modification should be done only to the numerator since the denominator term in (18) is insignificant in RAS design.

The numerator of (Eq. 2.18), i.e., $N_\Gamma(S) = \sum_{n=1}^{N} a_n S^n - 1$, is used to calculate the coefficients $a_1, a_2 \ldots a_n$. To find these coefficients for n layers, a standardized equation is needed. For this a term P_N is introduced, i.e.,

$$N_\Gamma(S) = \sum_{n=1}^{N} a_n S^n - 1 = (Z_c S + 1)P_N - P_{N+1} \tag{2.19}$$

The expression for P_N and the standardized equation to find the coefficients can be derived as follows (Toit 1993).

The ABCD matrix for n layers can be written as

$$\begin{bmatrix} A_n & B_n \\ C_n & D_n \end{bmatrix} = \begin{bmatrix} 1 & Z_c S \\ Y_n & Z_c Y_n S + 1 \end{bmatrix} \begin{bmatrix} A_{n-1} & B_{n-1} \\ C_{n-1} & D_{n-1} \end{bmatrix} \tag{2.20}$$

With $n = 2, 3, \ldots N$.

Thus, the general expression for B_n and D_n can be extracted as follows:

$$B_n = B_{n-1} + Z_c S D_{n-1} \tag{2.21}$$
$$D_n = Y_n B_{n-1} + (Z_c Y_n S + 1)D_{n-1} \tag{2.22}$$

where, for $n = 1$ from (Eq. 2.20), $B_1 = Z_c S$, and $D_1 = Z_c Y_1 S + 1$.

Substituting $n = n + 1$ in (Eq. 2.21), one gets

$$B_{n+1} = B_n + Z_c S D_n \tag{2.23}$$

Substituting value of D_n from (Eq. 2.22) in (Eq. 2.23), one has

$$\begin{aligned} B_{n+1} &= B_n + Z_c S[Y_n B_{n-1} + (Z_c Y_n S + 1)D_{n-1}] \\ &= B_n + Z_c S[Y_n B_{n-1} + Z_c S Y_n D_{n-1} + D_{n-1}] \\ &= B_n + Z_c S[Y_n B_{n-1} + Z_c Y_n D_{n-1} + D_{n-1}] \\ &= B_n + Y_n Z_c S B_{n-1} + Z_c^2 S^2 Y_n D_{n-1} + Z_c S D_{n-1} \\ &= B_n + (Z_c S Y_n)B_{n-1} + Z_c S(Z_c S Y_n + 1)D_{n-1} \end{aligned} \tag{2.24}$$

From (Eq. 2.21), $B_n = B_{n-1} + Z_c S D_{n-1}$

$$D_{n-1} = \frac{B_n - B_{n-1}}{Z_c S} \qquad (2.25)$$

Substituting (Eq. 2.25) in (Eq. 2.24),

$$B_{n+1} = B_n + (Z_c S Y_n) B_{n-1} + Z_c S (Z_c S Y_n + 1) \left[\frac{B_n - B_{n-1}}{Z_c S} \right]$$

On rearranging the equation,

$$B_{n+1} = B_n (2 + Z_c S Y_n) - B_{n-1} \qquad (2.26)$$

With $n = 1, 2, \ldots N - 1$, $B_0 = 0$, and $B_1 = Z_c S$.
Substituting (Eq. 2.25) in (Eq. 2.22), one gets

$$
\begin{aligned}
D_n &= Y_n B_{n-1} + (Z_c S Y_n + 1) \left(\frac{B_n - B_{n-1}}{Z_c S} \right) \\
&= Y_n B_{n-1} + \frac{Z_c S Y_n B_n - Z_c S Y_n B_{n-1} + B_n - B_{n-1}}{Z_c S} \\
&= \frac{Y_n B_{n-1} Z_c S + Z_c S Y_n B_n - Z_c S Y_n B_{n-1} + B_n - B_{n-1}}{Z_c S} \\
D_n &= \frac{1}{Z_c S} \{ (Z_c S Y_n + 1) B_n - B_{n-1} \} \quad \text{For } n = 2, 3, \ldots N
\end{aligned}
\qquad (2.27)
$$

From (Eq. 2.26),

$$B_{n+1} - B_n = B_n (Z_c S Y_n + 1) - B_{n-1}$$

Thus, (Eq. 2.27) can be written as

$$D_n = \frac{1}{Z_c S} \{ B_{n+1} - B_n \}, n = 1, 2, \ldots N \qquad (2.28)$$

$$\text{Defining,} \, P_n = \frac{1}{Z_c S} B_n, \quad n = 0, 1, 2, \ldots N + 1 \qquad (2.29)$$

The equation (Eq. 2.26) becomes

$$P_{n+1} = P_n (Z_c S Y_n + 2) - P_{n-1} \, \text{with} \, P_0 = 0, \, P_1 = 1, \, n = 1, 2, \ldots . N \qquad (2.30)$$

The equation (Eq. 2.18) for reflection coefficient can be re-written as

$$\Gamma(S) = \frac{Z_{in} - 1}{Z_{in} + 1} = \frac{B_N - D_N}{B_N + D_N} = \frac{B_N / D_N - 1}{B_N / D_N + 1} \qquad (2.31)$$

From (Eq. 2.28) and (Eq. 2.29),

$$B_N = Z_c S P_N \quad D_N = P_{N+1} - P_N$$

$$\Gamma(S) = \frac{\frac{Z_c S P_N}{P_{N+1} - P_N} - 1}{\frac{Z_c S P_N}{P_{N+1} - P_N} + 1} = \frac{(Z_c S + 1)P_N - P_{N+1}}{(Z_c S - 1)P_N + P_{N+1}} \tag{2.32}$$

The numerator of (Eq. 2.27) can be written as

$$N_\Gamma(S) = \sum_{n=1}^{N} a_n S^n - 1 = (Z_c S + 1)P_N - P_{N+1} \tag{2.33}$$

which is same as (Eq. 2.19).

In order to calculate the coefficients $a_1, a_2, \ldots a_N$, assume $P_0 = 0$ and $P_1 = 1$. One may notice that the numerator of reflection coefficient is expressed in terms of P_N and P_{N+1}. The equations for calculation of P_{N+1} and to find coefficients a_n are given in Appendix C.

For a single layer, there will be one coefficient, a_1. Likewise for a double-layer absorber, there will be two coefficients a_1 and a_2 corresponding to the admittance of each layer. Once a_n is known, the corresponding sheet admittance Y_n can be obtained. This gives the resistivity of the layer in the absorber.

As an example, let us consider a double-layer absorber, i.e., $N = 2$. Here the generalized equation is used to find the values of $a_1, a_2, \ldots a_N$ of two layers.

$$\text{The general equation is } P_n(S) = \sum_{m=0}^{n-1} p_m^{(n)} S^m, \text{ for } n$$

$$= 1, 2, \ldots N + 1 \tag{2.34}$$

For $N = 2$, it is required to find P_1, P_2, and P_3.
From (Eq. 2.34), $P_1(S) = p_0^{(1)} S^0 = 1$

$$P_2(S) = \sum_{m=0}^{1} p_m^{(2)} S^m = p_0^{(2)} S^0 + p_1^{(2)} S^1$$

$$P_3(S) = \sum_{m=0}^{2} p_m^{(3)} S^m = p_0^{(3)} S^0 + p_1^{(3)} S^1 + p_2^{(3)} S^2$$

Further, the expression (Eq. 2.34) should also satisfy the condition

$$p_m^{(n)} \equiv 0 \text{ for } m < 0 \text{ or } m \geq n \tag{2.35}$$

According to (Eq. 2.30), one has

$$P_{n+1} = P_n(Z_c S Y_n + 2) - P_{n-1}$$

Using (Eq. 2.34), the above equation can be written as

$$p_m^{(n)} = Z_c Y_{n-1} p_{m-1}^{(n-1)} + 2p_m^{(n-1)} - p_m^{(n-2)}, m = 0, 1, \ldots n-1 \qquad (2.36)$$

In this example, $m = 0, 1, 2$ and $n = 2, 3$.

Case 1 $n = 2$

$p_m^{(2)} = Z_c Y_1 p_{m-1}^1 + 2p_m^{(1)} - p_m^{(0)}$

For $n = 2$ and $m = 0$, (Eq. 2.36) can be written as

$p_0^{(2)} = Z_c Y_1 p_{-1}^1 + 2p_0^{(1)} - p_0^{(0)}$

Applying the condition (2.35), one gets

$p_0^{(2)} = 2p_0^1 = 2$

Continuing with same procedure, for $n = 2$ and $m = 1$

$p_1^{(2)} = Z_c Y_1 p_0^1 + 2p_1^{(1)} - p_1^{(0)} = Z_c Y_1$

For $n = 2$ and $m = 2$

$p_2^{(2)} = Z_c Y_1 p_1^1 + 2p_2^{(1)} - p_2^{(0)} = 0$

Case 2 $n = 3$

$p_0^{(3)} = Z_c Y_2 p_{m-1}^2 + 2p_m^{(2)} - p_m^{(1)}$

For $n = 3$ and $m = 0$

$p_0^{(3)} = Z_c Y_2 p_{-1}^2 + 2p_0^{(2)} - p_0^{(1)} = 2p_0^2 - p_0^1 = 3$

For $n = 3$ and $m = 1$

$p_1^{(3)} = Z_c Y_2 p_0^2 + 2p_1^{(2)} - p_1^{(1)} = 2(Z_c Y_2 + Z_c Y_1)$

For $n = 3$ and $m = 2$

$p_2^{(3)} = Z_c Y_2 p_1^2 + 2p_2^{(2)} - p_2^{(1)} = Z_c^2 Y_1 Y_2$

Now one has all the values for $p_m^{(n)}$.

Again in order to find out the coefficients a_1 and a_2 for two layers,

$$N_\Gamma(S) = \sum_{n=1}^{2} a_n S^n - 1$$

$$a_n = Z_c P_{n-1}^{(N)} + P_n^{(N)} - P_n^{(N+1)}, \quad \text{for } n = 1, 2$$

Therefore $a_1 = Z_c P_0^{(2)} + P_1^{(2)} - P_1^{(3)}$

$$a_1 = 2Z_c + Z_c Y_1 - (2Z_c Y_2 + 2Z_c Y_1) = Z_c(2 + Y_1 - 2Y_2 - 2Y_1)$$
$$= Z_c(2 - Y_1 - 2Y_2)$$
$$a_2 = Z_c P_1^{(2)} + P_2^{(2)} - P_2^{(3)} = Z_c^2 Y_1 + 0 - Z_c^2 Y_1 Y_2 = Z_c^2 Y_1(1 - Y_2)$$

Using these coefficients, admittance of each layer is determined. This provides the resistivity of each layer used in designing the absorber.

Table 2.3 summarizes the values of coefficients a_n up to three layers of the absorber (Toit 1993).

Table 2.3 Symbolic expansions $N_\Gamma(S)$ for $N = 1, 2, 3$

	$N = 1$	$N = 2$	$N = 3$
a_1	$Z_c(1 - Y_1)$	$Z_c(2 - Y_1 - 2Y_2)$	$Z_c(3 - Y_1 - 2Y_2 - 3Y_3)$
a_2	–	$Z_c^2 Y_1(1 - Y_2)$	$2Z_c^2(Y_1 + Y_2)(1 - Y_3) - Z_c^2 Y_1 Y_2$
a_3	–	–	$Z_c^3 Y_1 Y_2(1 - Y_3)$

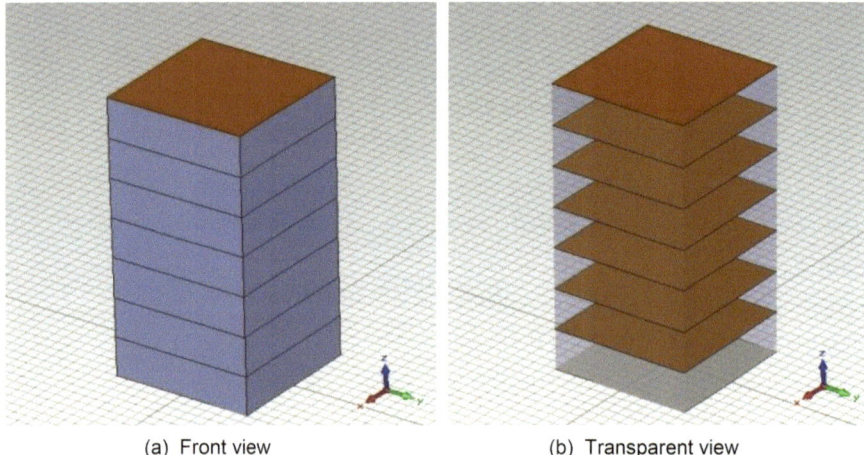

(a) Front view   (b) Transparent view

Fig. 2.10 Design of a seven-layer Jaumann absorber, (**a**) front view, (**b**) transparent view

Table 2.4 Dimension of absorber with seven layers

Parameters	Layer 1	Layer 2	Layer 3	Layer 4	Layer 5	Layer 6	Layer 7
Dimension of square patch (mm × mm)	6 × 6	6 × 6	6 × 6	6 × 6	6 × 6	6 × 6	6 × 6
Resistivity of square patch (Ω/Sq)	305	579	873	1266	1796	2480	3724
Width of dielectric (mm)	6	6	6	6	6	6	6
Thickness of dielectric (mm)	4	4	4	4	4	4	4

As an example, let us consider a seven-layer RAS structure. The design of a seven-layer Jaumann absorber is shown in Fig. 2.10. The parameters used for designing are tabularized in Table 2.4. Each layer of this absorber has a resistive sheet backed by a dielectric spacer ($\varepsilon_r = 1$).

The total thickness of the absorber depends on the operating frequency since each resistive sheet is separated by a distance $\lambda/4$. The reflection coefficient of design is shown in Fig. 2.11. It may be observed that the performance of the designed absorber is excellent both in terms of absorption and bandwidth.

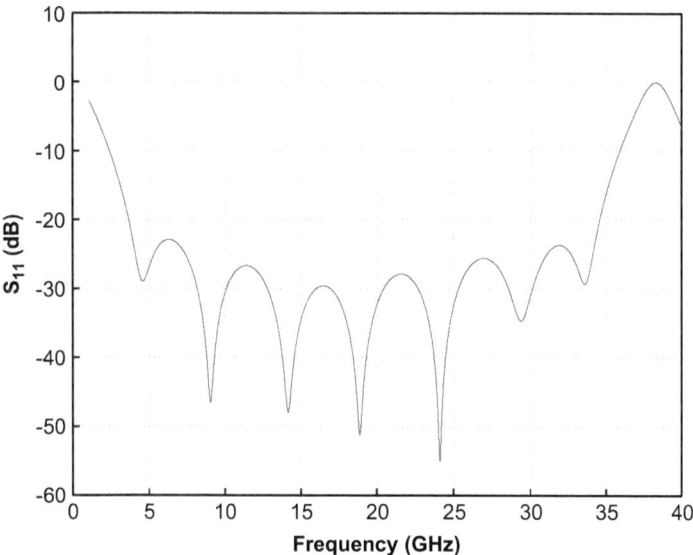

Fig. 2.11 Reflection coefficient of a seven-layer Jaumann absorber

References

Kazemzadeh A, Karlsson A (2009) Capacitive circuit method for fast and efficient design of wideband radar absorbers. IEEE Trans Antennas Propag 57(8):2307–2314

Lee, H.-M. and H.-S. Lee, "A Method for Extending the Bandwidth of Metamaterial Absorber," International Journal of Antennas and Propagation, vol. 2012 (2012), Article ID 859429, 7 pages. http://dx.doi.org/10.1155/2012/859429

Munk BA, Munk P, Prior J (2007) On designing Jaumann and Circuit Analog absorbers (CA Absorbers) for oblique angle of incidence. IEEE Trans Antennas Propag 55(1):186–192

Toit LJD (1993) Analysis and synthesis algorithms for the electric screen Jaumann electromagnetic wave Absorber, Ph.D. dissertation, Faculty of Engineering, University of Stellenbosch, South Africa, 75 p

Toit LJD (1994) The design of Jaumann absorbers. IEEE Anten Propag Mag 36(6):17–25

Chapter 3
Equivalent Circuit Model-Based RAS Design

The design of an absorber can be carried out based on its equivalent circuit. If the absorber consists of frequency selective surface (FSS), or metamaterials, or dielectric layer, the design can be done in terms of RLC components. An equivalent circuit helps the designer to analyze, design, and understand the absorption mechanism.

Figure 3.1 shows an N-layer absorber and its transmission line model. For designing an absorber, first the circuit components are to be determined. In the following section, it is explained how to calculate RLC values for a single-layer absorber in terms of effective impedance and reflection coefficient (Sjoiberg 2007).

3.1 Calculation of Effective Admittance

For an N-layer absorber, the relative admittance of the nth layer is given by

$$Y_n = Y_0 \sqrt{\varepsilon_n/\mu_n} \tag{3.1}$$

where Y_0 is the characteristic admittance and ε_n and μ_n are the permittivity and permeability of the nth layer, respectively..

For a nonmagnetic material, $\mu_n = 1$; therefore, (Eq. 3.1) becomes

$$Y_n = Y_0 \sqrt{\varepsilon_n} \tag{3.2}$$

In general, effective admittance Y_{eff} can be represented through a series RLC circuit (Costa et al. 2010):

© The Author(s) 2018
H. Singh et al., *Fundamentals of EM Design of Radar Absorbing Structures (RAS)*,
DOI 10.1007/978-981-10-5080-0_3

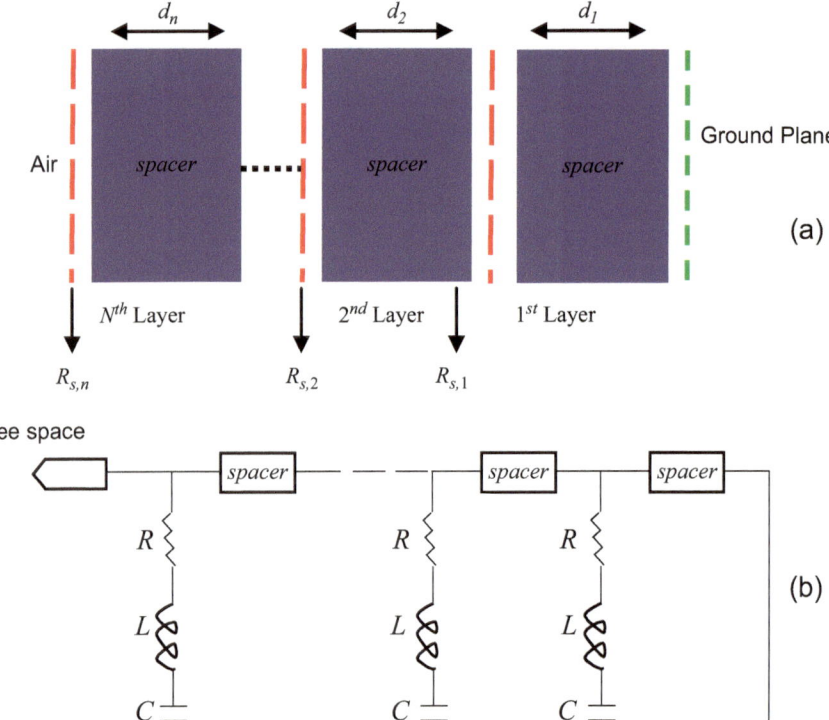

Fig. 3.1 (**a**) Schematic of an N-layer absorber. (**b**) Transmission line model

$$\frac{1}{Y} = Z_{\text{eff}} = R + j\omega L - j\frac{1}{\omega C} \tag{3.3}$$

where Z_{eff} is the effective impedance and R, L, C are the effective resistance, inductance, and capacitance, respectively. Z_{eff} and Y_{eff} are frequency dependent, whereas R, L, C are frequency independent. Moreover these parameters are related to the pattern geometry and sheet resistance of the layer (Chen et al. 2015).

3.2 Derivation of Circuit Element Parameters

Here a single-layer absorber is considered. A resistive sheet consisting of FSS is placed at a distance d above the ground plane. Figure 3.2 shows the transmission line equivalent of a single-layer absorber.

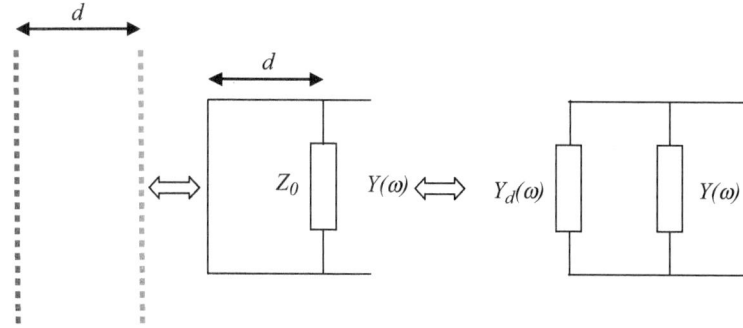

Fig. 3.2 Transmission line equivalent of a single-layer absorber

The admittance of transmission line of length d in the presence of ground plane is given by

$$Y_d(\omega) = -jY_0 \cot(kd\cos\theta) \tag{3.4}$$

At the distance $\lambda/4$ from the ground plane, one has

$$k_d d\cos\theta = \pi/2 \tag{3.5}$$

Using Taylor series expansion, (Eq. 3.4) becomes (Sjoberg 2007)

$$\begin{aligned} Y_d(\omega) &= -jY_0\left(0 - (k - k_d)d\cos\theta + O(k - k_d)^2\right) \\ &\approx \frac{jY_0(\omega - \omega_d)d\cos\theta}{c} \end{aligned} \tag{3.6}$$

Here, c is speed of light in free space.

In order to achieve good absorption over a broad frequency range, the admittance of an absorber should be equal to characteristic admittance, i.e., $Y(\omega) + Y_d(\omega) \approx Y_0$.

The admittance of the series modeled absorber sheet can be written as

$$\begin{aligned} Y(\omega) &= \frac{1}{R + j\omega L + \frac{1}{j\omega C}} = \frac{1/R}{1 + \frac{j}{R}\left[\omega L - \frac{1}{\omega C}\right]} \\ &= \frac{1/R}{1 + \frac{j}{R}\left[\omega\sqrt{LC} \times \frac{\sqrt{L}}{\sqrt{C}} - \frac{1}{\omega\sqrt{LC}} \times \frac{\sqrt{L}}{\sqrt{C}}\right]} = \frac{1/R}{1 + \frac{j}{R}\sqrt{\frac{L}{C}}\left[\omega\sqrt{LC} - \frac{1}{\omega\sqrt{LC}}\right]} \\ &= \frac{1/R}{1 + j\sqrt{\frac{L/C}{R}}\left[\frac{\omega}{1/\sqrt{LC}} - \frac{1/\sqrt{LC}}{\omega}\right]} \end{aligned} \tag{3.7}$$

Replacing $1/\sqrt{LC} = \omega_r$ the equation becomes

$$\frac{1/R}{1 + j\sqrt{\frac{L/C}{R}}\left[\frac{\omega}{\omega_r} - \frac{\omega_r}{\omega}\right]} = \frac{1/R}{1 + jQ\left[\frac{\omega}{\omega_r} - \frac{\omega_r}{\omega}\right]} = \frac{1/R}{1 + jQ\frac{\omega + \omega_r}{\omega_r\omega}(\omega - \omega_r)}$$

$$= \frac{1}{R} - j\frac{Q}{R}\frac{2}{\omega_r}(\omega - \omega_r) + O\left((\omega - \omega_r)^2\right) \tag{3.8}$$

where $Q = \sqrt{\frac{L/C}{R}}$ is the quality factor and $\omega_r = \frac{1}{\sqrt{LC}}$ is the resonant frequency and $\omega_d = k_d c$.

Choosing $\omega_r = \omega_d$, the reflection coefficient becomes real, i.e.,

$$\Gamma(\omega_r) = \frac{Y_0 - \{Y_d(\omega_r) + Y(\omega_r)\}}{Y_0 + \{Y_d(\omega_r) + Y(\omega_r)\}} = \frac{Y_0 - 1/R}{Y_0 + 1/R} \tag{3.9}$$

Taking $\Gamma(\omega_r) = -\Gamma_0$, where Γ_0 is a positive real number which gives the typical reflection level, one gets

$$-\Gamma_0 = \frac{Y_0 - 1/R}{Y_0 + 1/R} = \frac{Y_0 R - 1}{Y_0 R + 1}$$

$$-\Gamma_0 Y_0 R - \Gamma_0 = Y_0 R - 1$$

$$R(-\Gamma_0 Y_0 - Y_0) = \Gamma_0 - 1$$

$$R = \frac{\Gamma_0 - 1}{Y_0(-\Gamma_0 - 1)}$$

Therefore, the resistance R is given by

$$R = \frac{1}{Y_0}\left[\frac{1 - \Gamma_0}{1 + \Gamma_0}\right] \tag{3.10}$$

$Y(\omega) + Y_d(\omega) = \frac{1}{R} - j\frac{Q}{R}\frac{2}{\omega_r}(\omega - \omega_r) + O\left((\omega - \omega_r)^2\right) + \frac{jY_0(\omega - \omega_d)d\cos\theta}{c}$ Now, equating imaginary part of $Y(\omega) + Y_d(\omega)$ at first order in $(\omega - \omega_r)$, to zero, one gets

$$\frac{Y_0 d\cos\theta}{c} = \frac{2Q}{R\omega_r} = \frac{2\sqrt{L/C}/R}{R/\sqrt{LC}} = \frac{2L}{R^2} \tag{3.11}$$

Hence the inductance is obtained as

$$L = \frac{R^2 Y_0 c^{-1} d\cos\theta}{2} = \frac{d\cos\theta}{2Y_0 c}\left[\frac{1 - \Gamma_0}{1 + \Gamma_0}\right]^2 \tag{3.12}$$

At $\omega_r = \omega_d$, $\omega_d = k_d c = \frac{\pi/2}{d\cos\theta}c$ from (Eq. 3.5)

$$\frac{1}{\sqrt{LC}} = \frac{\pi/2}{c^{-1}d\cos\theta}, LC = \left(\frac{c^{-1}d\cos\theta}{\pi/2}\right)^2$$

$$C = \frac{1}{L}\left(\frac{c^{-1}d\cos\theta}{\pi/2}\right)^2 = \left(\frac{1}{\frac{c^{-1}d\cos\theta}{2Y_0}\left[\frac{1-\Gamma_0}{1+\Gamma_0}\right]^2}\right)\left(\left(\frac{c^{-1}d\cos\theta}{\pi/2}\right)^2\right)$$

Hence capacitance can be calculated as

$$C = \frac{4}{\pi^2}\frac{2Y_0d\cos\theta}{c}\left[\frac{1+\Gamma_0}{1-\Gamma_0}\right]^2 \tag{3.13}$$

Hence, the general equations for R, L, and C of a single-layer absorber are derived as

$$R = \frac{1}{Y_0}{}_0\left[\frac{1-\Gamma_0}{1+\Gamma_0}\right]$$

$$L = \frac{d\cos\theta}{2Y_0c}\left[\frac{1-\Gamma_0}{1+\Gamma_0}\right]^2$$

$$C = \frac{4}{\pi^2}\frac{2Y_0d\cos\theta}{c}\left[\frac{1+\Gamma_0}{1-\Gamma_0}\right]^2$$

Let us consider one example of a single-layer absorber of $\lambda/4$ thickness at 4.77 GHz. The equivalent circuit of the absorber is modeled as a series RLC and a transmission line parallel to the RLC on one side and short circuited at the other end (Fig. 3.3).

Assume that the reflection level to be obtained is below -20 dB for normal incidence. Accordingly reflection coefficient becomes 0.1, $k_d d\cos\theta = \pi/2$, $\lambda = 0.0628$ m for 4.77 GHz, and $k_d = k_0\cos\theta = 2\pi/\lambda$ for $\cos\theta = 1$. The corresponding effective resistance, effective inductance, and effective capacitance can be calculated using (Eq. 3.10), (Eq. 3.12), and (Eq. 3.13), respectively.

$$R = \frac{1}{Y_0}{}_0\left[\frac{1-\Gamma_0}{1+\Gamma_0}\right]$$

On substituting the values,

$$R = 377\left[\frac{1-0.1}{1+0.1}\right] = 308.45\ \Omega$$

Fig. 3.3 Equivalent circuit
of a single-layer absorber

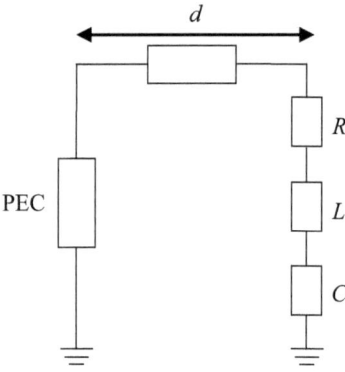

Inductance is

$$L = \frac{d \cos \theta}{2Y_0 c} \left[\frac{1 - \Gamma_0}{1 + \Gamma_0} \right]^2$$

$$L = \frac{\pi/2k_d}{2 \times 2.66 \times 10^{-3} \times 3 \times 10^8} \left[\frac{1 - 0.1}{1 + 0.1} \right]^2, k_d = \frac{2\pi}{\lambda}$$

Substituting the values,

$$L = \frac{\lambda/4}{1596000} \times 0.6694 = 6.58 \text{ nH}$$

Capacitance is

$$C = \frac{4}{\pi^2} \frac{2Y_0 d \cos \theta}{c} \left[\frac{1 + \Gamma_0}{1 - \Gamma_0} \right]^2$$

$$C = \frac{2 \times 2.66 \times 10^{-3} \times \lambda/4 \times 2\pi/k_d \lambda}{3 \times 10^8 \times \pi^2/4} \left[\frac{1 + 0.1}{1 - 0.1} \right]^2$$

$$= \frac{2 \times 2.66 \times 10^{-3} 2 \times 0.0628 \times 1.493}{3 \times 10^8 \times 2\pi^2} = 0.16 \text{pF}$$

Once, R, L, and C are known, the effective admittance can be calculated for proceeding toward EM design of the absorber with required absorption behavior.

The operating frequency range of the above absorber is assumed to be 2.77 GHz – 6.77 GHz. Thus, 4.77 GHz is the center frequency. From (Eq. 3.3), the effective impedance is obtained as

$$Z_{eff} = R + j\omega L - j\frac{1}{\omega C}$$

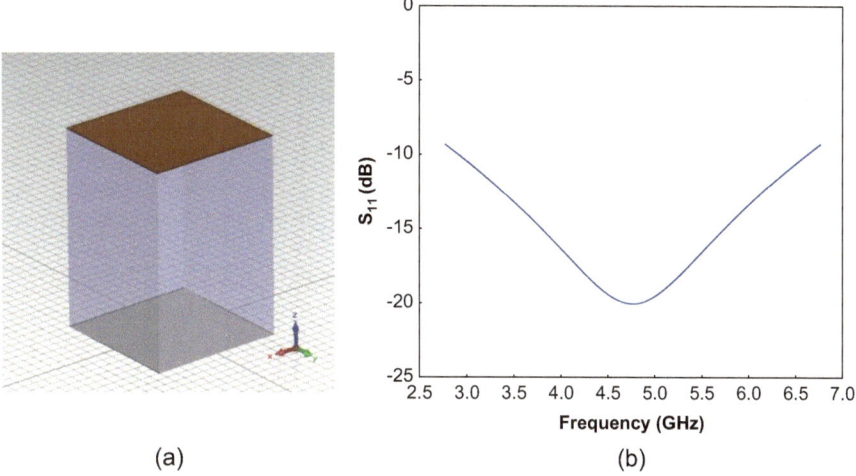

Fig. 3.4 Reflection behavior of a single-layer absorber, (**a**) design, (**b**) reflection coefficient

$$Z_{\text{eff}} = R + \left[\frac{1 - \omega^2 LC}{j\omega C}\right] = 308.45 + \left[\frac{0.0543}{j4.7952 \times 10^{-3}}\right] = 308.45 - 11.323j\,\Omega$$

Thus, $|Z_{\text{eff}}| = 308.6577\ \Omega$.

This effective impedance is then normalized w.r.t. that of free space impedance. This gives normalized impedance of 0.8169. The corresponding effective admittance is 1.2240.

The resistivity of the sheet is calculated as

$$R_{s,n} = \frac{\eta}{Y_{\text{eff}}} = \frac{377}{1.22402} = 308\ \Omega/\text{Sq}$$

When full-wave simulation is carried out using this resistivity value, the reflection behavior of the design obtained is shown in Fig. 3.4. The distance of the sheet from ground plane is $d = 1.57$ cm (i.e., $\lambda/4$). This reflection characteristic can be improved by using FSS-based resistive sheet or adding more layers in the design.

References

Chen M, Zhu X, Lei H, Fang D, Zhang Z (2015) Optimal design of broadband radar absorbing sandwich structure with circuit analog absorber core. Int J Appl Mech 7(2):13

Costa F, Monorchio A, Manara G (2010) Analysis and Design of Ultra Thin electromagnetic absorbers comprising resistively loaded high impedance surfaces. IEEE Trans Antennas Propag 58(5):1551–1558

Sjoiberg D (2007) Circuit analogs for stratified structures, Report, CODEN:LUTEDX/(TEAT-7159), 18p

Chapter 4
Smith Chart-Based RAS Design

Smith chart is one of the graphical tools used in microwave engineering. It can be used as a reflection coefficient to impedance/admittance converter or vice versa $(\Gamma \Leftrightarrow Z, Y)$. It is a valuable tool for the analysis of complex design problems. It provides help in finding impedance matching solutions avoiding lengthy calculations. It can be also used as a method to test whether the desired bandwidth can be attained practically. One can analyze the circuit performance using Smith chart, when circuit parameters change or new elements are added in the circuit (Radmanesh 2002).

The EM design of absorbers involves complex calculations. For better understanding and reduced complexity, Smith chart can be used. The first stage for this approach is to transform the structure into a transmission line model in which the transmission line is shorted at one end and loaded by an impedance Z_L at $\lambda/4$ distance from the short circuit end (Elia et al. 2010), as shown in Fig. 4.1.

The short circuit becomes open circuit at the distance of $\lambda/4$, and this is true only for normal incidence. The details of steps involved are given in Appendix D. While designing an absorber, the load impedance Z_L should be matched with that of air Z_{Air} for perfect absorption at the desired frequency.

4.1 Single-Layer Absorber

A single-layer absorber also referred to as the Salisbury screen is shown in Fig. 4.2. It consists of a resistive sheet, a metal substrate, and a dielectric layer. The resistive sheet is mounted one quarter wavelength in front of a ground plane (Munk 2000).

The metal substrate is considered as short circuit and is marked at the short circuit point $(Z = 0)$ of Smith chart. For a dielectric layer of $\lambda/4$ thickness, the short circuit impedance is rotated by 180° to reach the open circuit impedance point $(Z = \infty)$. Next, a parallel resistive sheet with impedance equal to that of free

© The Author(s) 2018
H. Singh et al., *Fundamentals of EM Design of Radar Absorbing Structures (RAS)*,
DOI 10.1007/978-981-10-5080-0_4

Fig. 4.1 Equivalent
transmission line model of
generic absorber

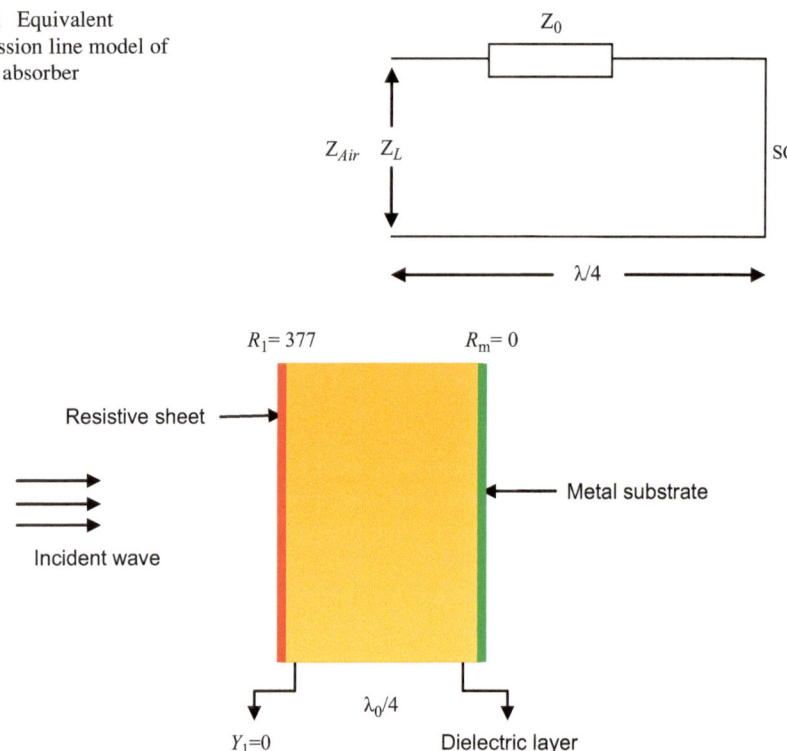

Fig. 4.2 Salisbury screen

Fig. 4.3 Transmission line
model of Salisbury screen

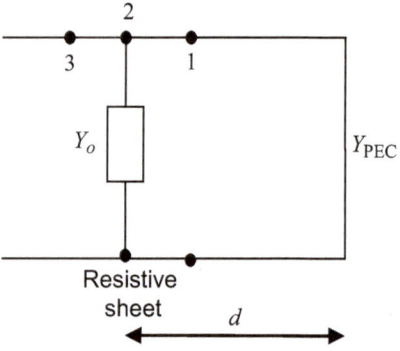

space is added. This gives zero reflection point on the Smith chart, i.e., center of
the Smith chart (Ke et al. 2011). In order to analyze this simplest design using
Smith chart, the structure is transformed into equivalent transmission line model,
shown in Fig. 4.3.

Table 4.1 Admittance and impedance values for ground plane

Frequency (GHz)	Admittance (Y_g)	Impedance (Z_g)
4	$-1.3764i$	$0.7265i$
5	$-1.0000i$	$1.0000i$
6	$-0.7265i$	$1.3764i$
7	$-0.5095i$	$1.9626i$
8	$-0.3249i$	$3.0770i$
9	$-0.1584i$	$6.3138i$
10	0	∞
11	$0.1584i$	$-6.3138i$
12	$0.3249i$	$-3.0770i$
13	$0.5095i$	$-1.9626i$
14	$0.7265i$	$-1.3764i$
15	$1.0000i$	$-1.0000i$
16	$1.3764i$	$-0.7265i$

The operating frequency of the above absorber is assumed to be 4–16 GHz. The spacing $d = 0.75$ cm, i.e., $\lambda/4$ at center frequency of 10 GHz. The transmission line model for the absorber is solved from the ground plane admittance Y_{PEC} to free space admittance Y_0 through resistive sheet. This resistive sheet can be a FSS sheet as per application.

First step in this approach is to normalize the Smith chart to free space admittance Y_0. Next step is to find the admittance at Point 1, i.e., ground plane admittance Y_{PEC}. The ground plane admittance Y_{PEC} is given by (Xu et al. 2014)

$$Y_{PEC} = \frac{-j \cot \left(2\pi d \sqrt{\varepsilon_r}/\lambda \right)}{Z_c} \tag{4.1}$$

Here Z_c and ε_r are the characteristic impedance and relative permittivity of PEC ground plane. λ represents the wavelength of the incident wave and $d = \lambda/4$. The admittance Y_{PEC} for the entire frequency band from lower frequency (f_L) to higher frequency (f_H) is calculated using (Eq. 4.1). The calculated values of admittance of ground plane are shown in Table 4.1. These admittance values will be located at the rim of the Smith chart, i.e., Curve 1, as shown in Fig. 4.4.

Next step is to find the Point 2 representing sheet admittance Y_0 as shown in Fig. 4.5. This corresponds to zero reflection, a point close to the center of the Smith chart. Then this admittance point should be added to curve 1, i.e., Y_{PEC} to obtain Curve 3, as shown in Fig. 4.6.

The Curve 3 would provide admittance for the entire frequency range (4–16 GHz). These admittance values would provide insight of reflection behavior directly from the Smith chart, avoiding full-wave analysis.

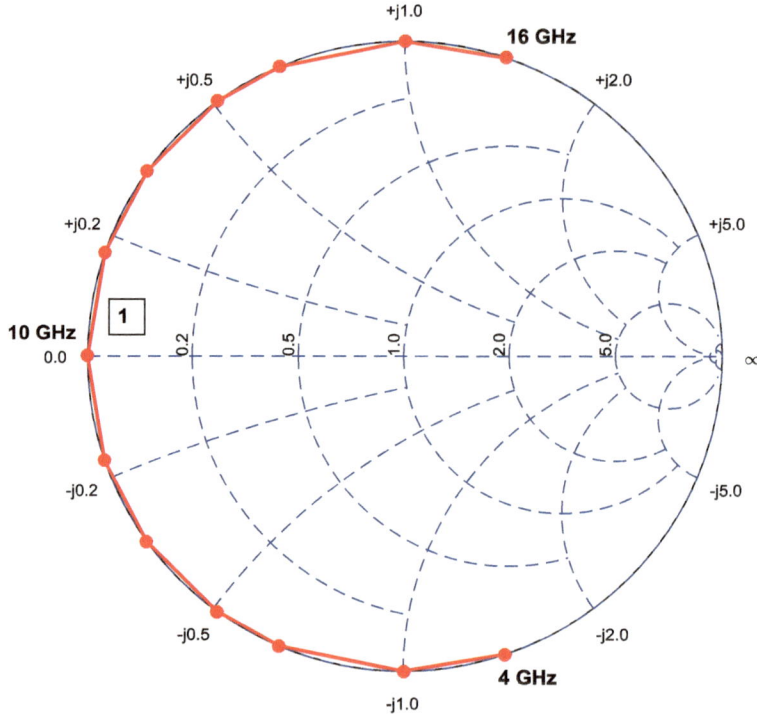

Fig. 4.4 Admittance curve as a function of frequency for ground plane (Point 1)

4.2 Double-Layer Absorber

The bandwidth obtained using single layer is very less. Moreover a single resistive sheet with admittance close to that of free space is not, in general, available. Thus, for wider absorption bandwidth, compensating dielectric matching plane (d_2) is added in front of the resistive sheet. The schematic of such a structure is shown in Fig. 4.7.

The admittance at Point 1 is the ground plane admittance, and it is purely imaginary as shown in Table 4.1. This point lies on the rim of the Smith chart. Curve 1, corresponding to the entire frequency range (4–16 GHz), is shown in Fig. 4.8. The impedance values are obtained using (Eq. 4.1).

Above ground plane, a dielectric spacer is added with thickness $d_1 = 7.49$ mm ($\lambda_1/4$). Since the admittance of this dielectric is same as ground plane, Curve 1 is valid on Smith chart. Next a resistive sheet with admittance of $1.7Y_0$ is placed, represented by Point 2 in the schematic (Fig. 4.7). In the Smith chart, it is shown as Point 2 (Fig. 4.9). Over this resistive sheet, one compensating dielectric sheet ($\varepsilon_r = 1.7$) is added with thickness $d_2 = 5.74$ mm ($\lambda_2/4$). This is represented by Point 3 in Fig. 4.7. The admittance at Point 3 is obtained by summing the Curve

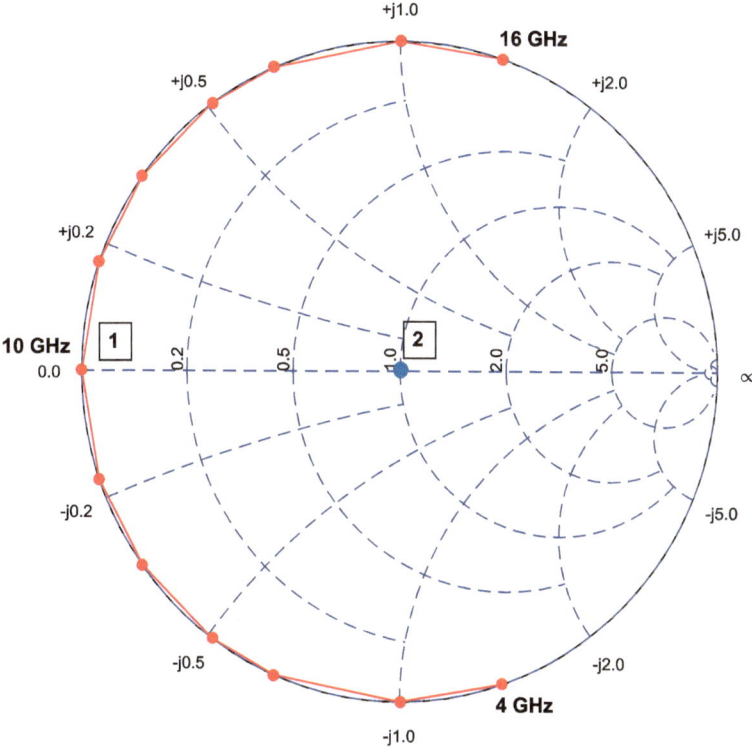

Fig. 4.5 Admittance curve as a function of frequency for resistive sheet (Point 2)

1 and Point 2 in Smith chart. This results in Curve 3 as shown in Fig. 4.10. Finally the admittance at Point 4 is obtained by traveling through the dielectric slab d_2 shown in Fig. 4.7. The admittance of Curve 4 at center frequency is found by rotating $\lambda_2/4$ distance (Fig. 4.11). The rotation is more for higher frequencies more than $\lambda_2/4$ is rotated and less than $\lambda_2/4$ is rotated at lower frequencies.

The reflection behavior of the structure is shown in Fig. 4.12. It is obtained by performing full-wave analysis. It is to be noted that the values of reflection coefficient over the frequency range (4 GHz – 16 GHz) can also be obtained directly from Smith chart as per Curve 4. It may be observed that the structure shows good absorption over a wide bandwidth. Thus, one may infer that the input impedance varies with increase of layers and contributes toward improvement in reflection characteristic at desired frequency.

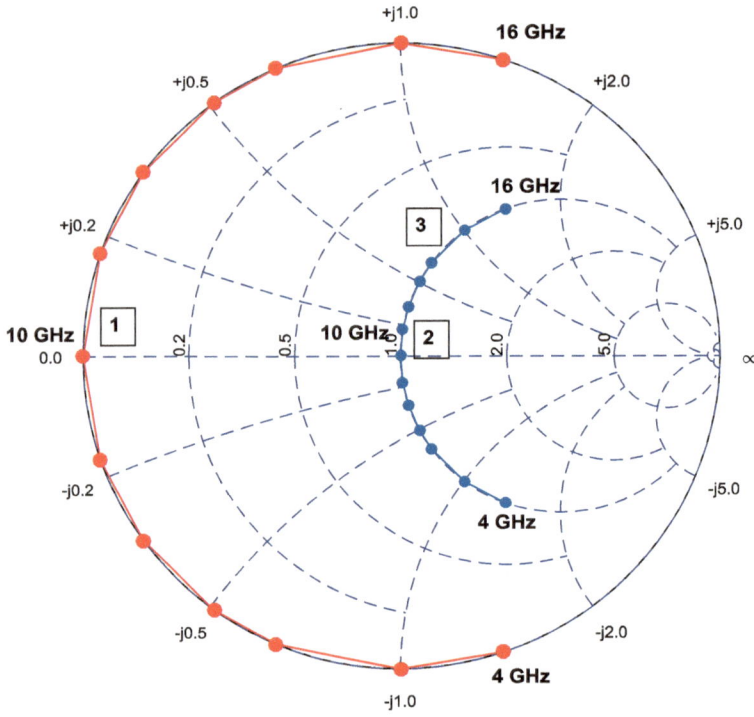

Fig. 4.6 Admittance curve as a function of frequency for entire structure (Point 3)

Fig. 4.7 Schematic of an absorber with dielectric slab d_2 in front of resistive sheet

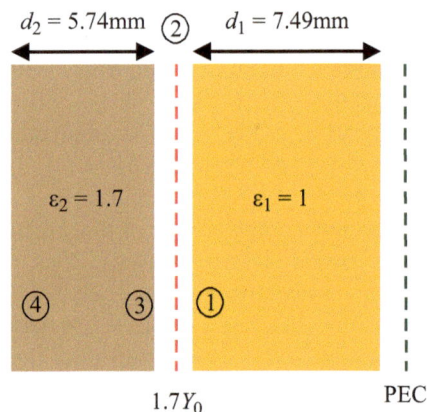

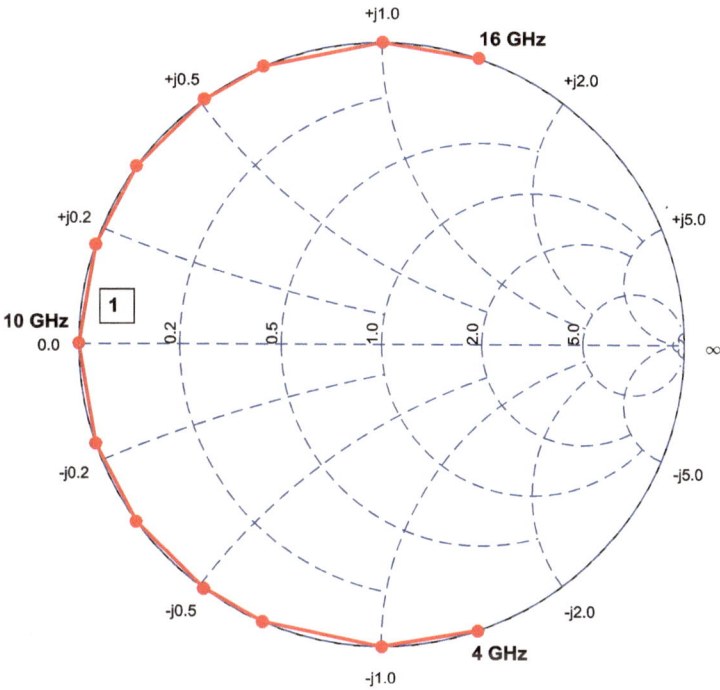

Fig. 4.8 Admittance curve as a function of frequency for ground plane

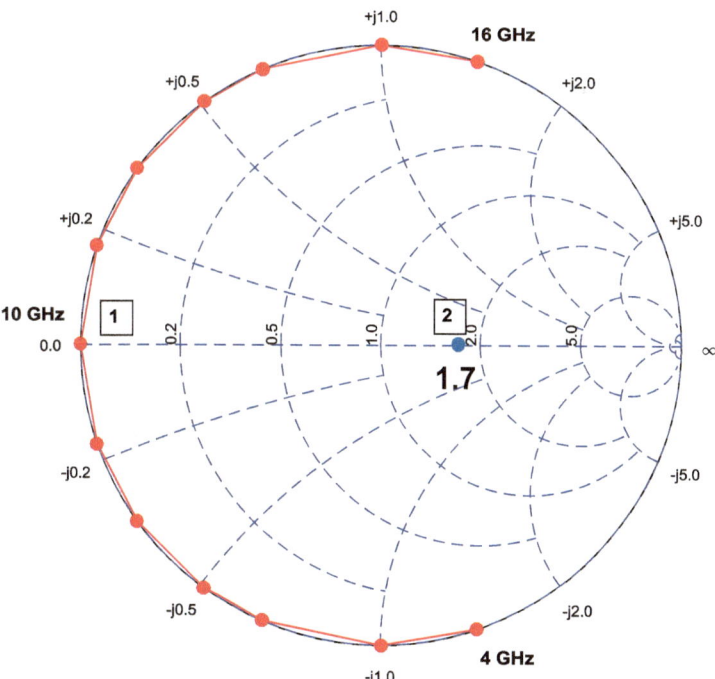

Fig. 4.9 Admittance curve as a function of frequency for ground plane and resistive sheet (Point 2)

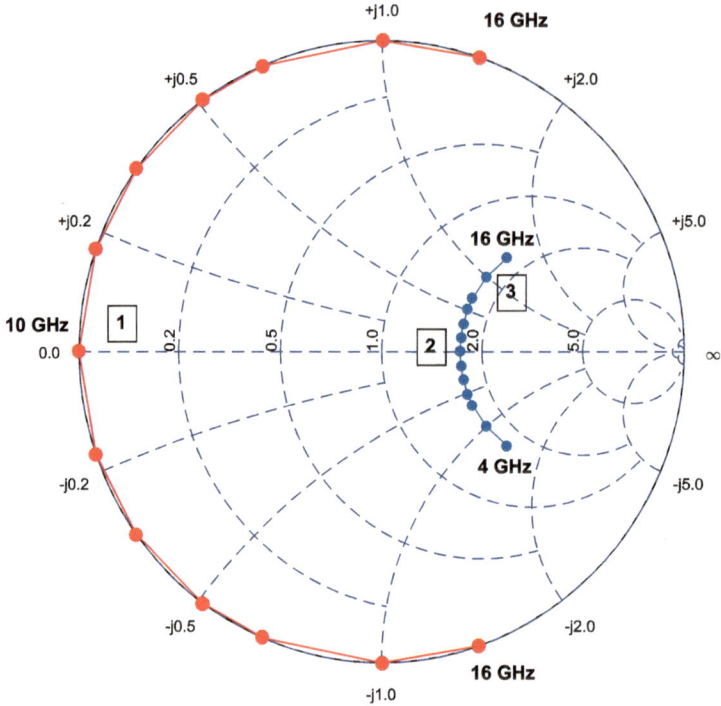

Fig. 4.10 Admittance curve as a function of frequency for ground plane and resistive sheets (Point 3)

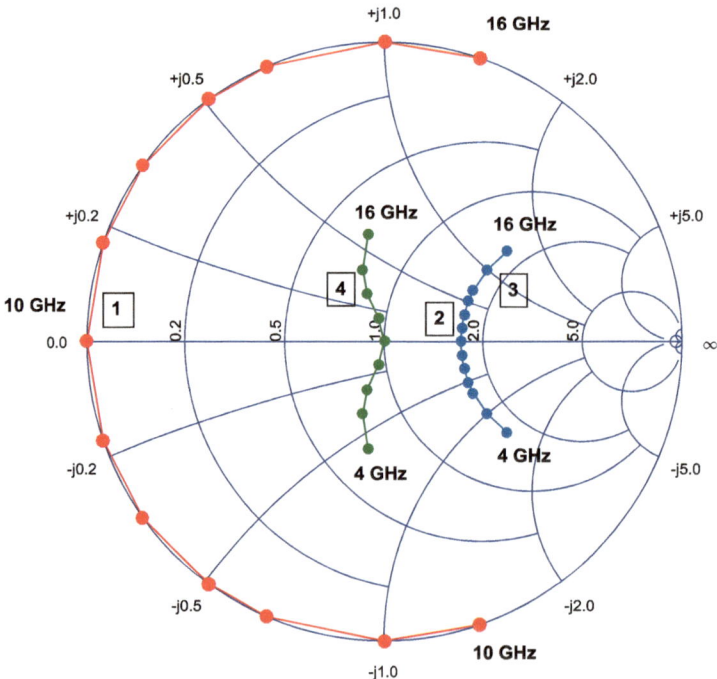

Fig. 4.11 Admittance curve as a function of frequency for full structure (Point 4)

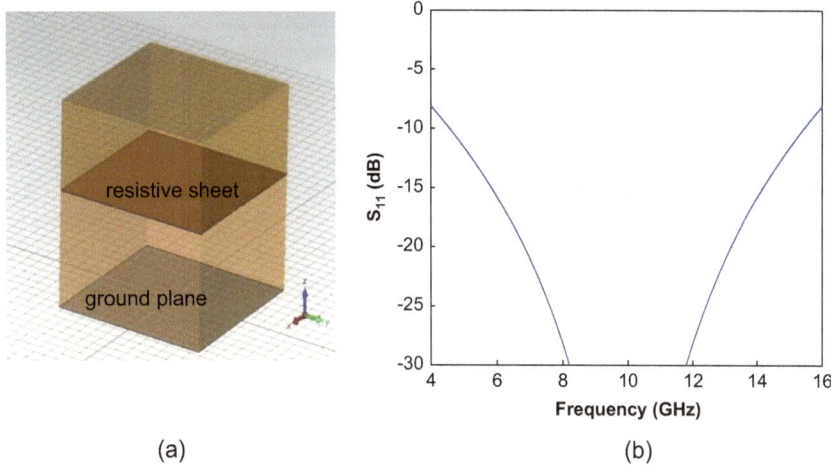

(a)   (b)

Fig. 4.12 Reflection behavior of full structure shown in Fig. 4.6. (**a**) design, (**b**) reflection coefficient

References

Elia UFD, Pelosi G, Selleri S, Taddei R (2010) A carbon-nanotube-based-frequency-selective absorber. Int J Microw Wirel Technol 2(5):479–485

Ke L, Xin Z, Xinyu H, Peng Z (2011) Analysis and design of multilayer Jaumann absorbers, In: Proceedings of International Conference on Microwave Technology and Computational Electromagnetics (ICMTCE), Beijing, pp 81–84, 22–25

Munk BA (2000) Frequency selective surfaces: theory and design. Wiley, New York, ISBN: 0-471-37047-9, 410 p

Radmanesh MM (2002) Radio Frequency and microwave electronics illustrated. Asia: Pearson Education, ISBN: 81-7808-284-5, 849 p

Xu X, Wang Q, Tang Z, Sun B (2014) Optimal design of non-magnetic metamaterial absorbers using visualization method. IEICE Electron Express 11(18):7

Chapter 5
Conclusion

The design of RAS involves calculation of circuit parameters such as R, L, C, Y, Z, etc. before proceeding toward full-wave analysis. These parameters depend on the geometry and frequency band considered. This document attempts to explain the fundamentals of RAS design using different approaches such as equivalent circuit approach and Smith chart. It is discussed that, in order to achieve the desired absorption behavior and bandwidth of RAS, a designer has to take into account different factors.

© The Author(s) 2018   37
H. Singh et al., *Fundamentals of EM Design of Radar Absorbing Structures (RAS)*,
DOI 10.1007/978-981-10-5080-0_5

Appendices

Appendix A: ABCD Voltage Matrix Analysis

Consider a single section of Jaumann absorber or Salisbury screen (Fig. A.1). Z_c is characteristic impedance of unit element, and S is the corresponding frequency element.

Applying node and mesh analysis,

$$V_1 = \frac{I_1 + I_2}{Y_n} \tag{A.1}$$

$$I_1 = V_1 Y_n - \frac{V_2 - V_1}{Z_c S} \tag{A.2}$$

Substituting (Eq. A.2) in (Eq. A.1), one gets

$$V_1 = \frac{V_1 Y_n Z_c S - V_2 + V_1 + I_2 Z_c S}{Z_c S Y_n}$$

$$V_1 Z_c S Y_n = V_1 Z_c S Y_n - V_2 + V_1 + I_2 Z_c S$$

$$V_2 - V_1 - I_2 Z_c S = 0$$

$$\text{i.e., } V_1 = V_2 - I_2 Z_c S \tag{A.3}$$

Comparing the above equation with

$$\begin{bmatrix} V_1 \\ I_1 \end{bmatrix} = \begin{bmatrix} A & B \\ C & D \end{bmatrix} \begin{bmatrix} V_2 \\ -I_2 \end{bmatrix}$$

one has $A = 1$ and $B = Z_c S$.

Substituting (Eq. A.3) in (Eq. A.2),

© The Author(s) 2018
H. Singh et al., *Fundamentals of EM Design of Radar Absorbing Structures (RAS)*,
DOI 10.1007/978-981-10-5080-0

Fig. A.1 One section of
Jaumann absorber

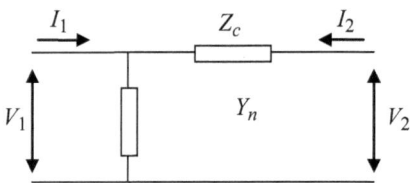

$$I_1 = (V_2 - I_2 Z_c S) Y_n - \frac{V_2 - (V_2 - I_2 Z_c S)}{Z_c S}$$

$$I_1 = \frac{V_2 Y_n Z_c S - I_2 (Z_c S)^2 Y_n - I_2 Z_c S}{Z_c S}$$

$$I_1 = V_2 Y_n - I_2 Z_c S Y_n - I_2,$$

Therefore I_1 becomes

$$I_1 = V_2 Y_n - I_2 (Z_c S Y_n + 1) \tag{A.4}$$

Then comparing with ABCD voltage matrix, one gets

$$C = Y_n \text{ and } D = Z_c S Y_n + 1$$

Hence the ABCD voltage matrix for one section of Jaumann absorber can be written as

$$\begin{bmatrix} V_1 \\ I_1 \end{bmatrix} = \begin{bmatrix} 1 & Z_c S \\ Y_n & Z_c S Y_n + 1 \end{bmatrix} \begin{bmatrix} V_2 \\ -I_2 \end{bmatrix} \tag{A.5}$$

The derived ABCD matrix can be validated using the fact that it satisfies the condition for reciprocity:

$$AD - BC = 1$$

Appendix B: Derivation of Admittance of Absorber Layers

The general equation to find the admittance values Y_1 and Y_2 for a two-layer cascaded transmission matrix is derived in this appendix.

The coefficients of Γ for two-layer Jaumann absorber are expressed as

$$a_1 = Z_c^2 Y_1 (1 - Y_2) \tag{B.1}$$

$$a_2 = Z_c(2 - Y_1 - 2Y_2) \tag{B.2}$$

Calculation of Y_1

From (Eq. B.2), one can write as

$$\frac{a_2}{Z_c} = 2 - Y_1 - 2Y_2$$

$$2Y_2 = 2 - Y_1 - \frac{a_2}{Z_c} \tag{B.3}$$

$$2Y_2 = \frac{2Z_c - Y_1 Z_c - a_2}{Z_c}$$

$$Y_2 = \frac{2Z_c - Y_1 Z_c - a_2}{2Z_c} \tag{B.4}$$

Substituting (Eq. B.4) in (Eq. B.1),

$$a_1 = Z_c^2 Y_1 - Z_c^2 Y_1 \left(\frac{2Z_c - Y_1 Z_c - a_2}{2Z_c}\right)$$

$$a_1 = \frac{2Z_c^2 Y_1 - Z_c Y_1(2Z_c - Y_1 Z_c - a_2)}{2} = \frac{2Z_c^2 Y_1 - 2Z_c^2 Y_1 + Z_c^2 Y_1^2 + a_2 Z_c Y_1}{2}$$

$$a_1 = \frac{Z_c^2 Y_1^2 + a_2 Z_c Y_1}{2}$$

$$2a_1 = Z_c^2 Y_1^2 + a_2 Z_c Y_1$$

$$Z_c^2 Y_1^2 + a_2 Z_c Y_1 - 2a_1 = 0,$$

Then from general quadratic equation form, i.e., $ax^2 + bx + c = 0$, one gets

$$a = Z_c^2; \quad b = a_2 Z_c; \quad c = -2a_1$$

Solving for Y_1,

$$Y_1 = \frac{-a_2 \pm \sqrt{\Delta}}{2Z_c} \qquad \Delta = b^2 + 8a_1 \tag{B.5}$$

Calculation of Y_2

From (Eq. B.3), one has

$$2Y_2 = 2 - Y_1 - \frac{a_2}{Z_c}$$

$$2Y_2 = 2 - \left[\frac{-a_2 + \sqrt{\Delta}}{2Z_c} \right] - \frac{a_2}{Z_c}$$

$$2Y_2 = 2 + \frac{a_2}{2Z_c} - \frac{\sqrt{\Delta}}{2Z_c} - \frac{a_2}{Z_c}$$

$$2Y_2 = 2 - \frac{a_2}{2Z_c} - \frac{\sqrt{\Delta}}{2Z_c}$$

Dividing by 2, one has

$$Y_2 = 1 + \frac{-a_2 \pm \sqrt{\Delta}}{4Z_c} \tag{B.6}$$

Therefore, the solution for the admittance can be written as

$$\text{Solution A}: \begin{cases} Y_1 = \dfrac{-a_2 + \sqrt{\Delta}}{2Z_c} \\ Y_2 = 1 + \dfrac{-a_2 - \sqrt{\Delta}}{4Z_c} \end{cases}$$

$$\text{Solution B}: \begin{cases} Y_1 = \dfrac{-a_2 - \sqrt{\Delta}}{2Z_c} \\ Y_2 = 1 + \dfrac{-a_2 + \sqrt{\Delta}}{4Z_c} \end{cases} \quad \text{where } \Delta = a_2^2 + 8a_1$$

Appendix C: Derivation of $N_\Gamma(S)$ of N-Layer Absorber

The ABCD matrix for n layers can be written as

$$\begin{bmatrix} A_n & B_n \\ C_n & D_n \end{bmatrix} = \begin{bmatrix} 1 & Z_c S \\ Y_n & Z_c Y_n S + 1 \end{bmatrix} \begin{bmatrix} A_{n-1} & B_{n-1} \\ C_{n-1} & D_{n-1} \end{bmatrix} \tag{C.1}$$

With $n = 2, 3, \ldots N$.
Thus, the general expression for B_n and D_n can be extracted as

$$B_n = B_{n-1} + Z_c S D_{n-1} \tag{C.2}$$

$$D_n = Y_n B_{n-1} + (Z_c Y_n S + 1) D_{n-1} \tag{C.3}$$

where, for $n=1$ from (Eq. C.1), $B_1 = Z_c S$, and $D_1 = Z_c Y_1 S + 1$.
Transforming $n \to n+1$ in (Eq. C.2), one gets

$$B_{n+1} = B_n + Z_c S D_n \tag{C.4}$$

Substituting value of D_n in (Eq. C.4), one has

$$
\begin{aligned}
B_{n+1} &= B_n + Z_c S[Y_n B_{n-1} + (Z_c Y_n S + 1)D_{n-1}] \\
&= B_n + Z_c S[Y_n B_{n-1} + Z_c Y_n S D_{n-1} + D_{n-1}] \\
&= B_n + (Z_c S Y_n)B_{n-1} + Z_c S(Z_c S Y_n + 1)D_{n-1}
\end{aligned} \tag{C.5}
$$

From (Eq. C.2), $B_n = B_{n-1} + Z_c S D_{n-1}$

$$D_{n-1} = \frac{B_n - B_{n-1}}{Z_c S} \tag{C.6}$$

Substituting (Eq. C.6) in (Eq. C.5),

$$B_{n+1} = B_n + (Z_c S Y_n)B_{n-1} + Z_c S(Z_c S Y_n + 1)\left[\frac{B_n - B_{n-1}}{Z_c S}\right]$$

On rearranging the equation, one gets

$$B_{n+1} = B_n(2 + Z_c S Y_n) - B_{n-1} \tag{C.7}$$

With $n = 1, 2, \ldots N - 1$, $B_0 = 0$, and $B_1 = Z_c S$.
Substituting (Eq. C.6) in (Eq. C.3), one gets

$$
\begin{aligned}
D_n &= Y_n B_{n-1} + (Z_c S Y_n + 1)\left(\frac{B_n - B_{n-1}}{Z_c S}\right) \\
&= Y_n B_{n-1} + \frac{Z_c S Y_n B_n - Z_c S Y_n B_{n-1} + B_n - B_{n-1}}{Z_c S} \\
&= \frac{Z_c S Y_n B_{n-1} + Z_c S Y_n B_n - Z_c S Y_n B_{n-1} + B_n - B_{n-1}}{Z_c S}
\end{aligned}
$$

$$D_n = \frac{1}{Z_c S}\{(Z_c S Y_n + 1)B_n - B_{n-1}\} \quad \text{For } n = 2, 3, \ldots N \tag{C.8}$$

From (Eq. C.7),

$$B_{n+1} - B_n = B_n(Z_c S Y_n + 1) - B_{n-1}$$

Thus, (Eq. C.8) can be written as

$$D_n = \frac{1}{Z_c S}\{B_{n+1} - B_n\} \ , \quad n = 2, 3, \ldots N \tag{C.9}$$

$$\text{Defining,} \quad P_n = \frac{1}{Z_cS}B_n, \quad n = 0, 1, 2, \ldots N+1 \tag{C.10}$$

The equation (Eq. C.7) becomes

$$P_{n+1} = P_n(Z_cSY_n + 2) - P_{n-1}, \quad \text{with} \quad P_0 = 0, P_1 = 1, n = 1, 2, \ldots N \tag{C.11}$$

This series of polynomials $P_0 \ldots \ldots \ldots \ldots _{N+1}(S)$ forms the basic building block of most of the analysis of N-layered radar absorber.

Computation of Coefficients of $P_n(S)$

To estimate the admittance of N layer, it is required to find the polynomial coefficients:

$$P_n(S) = \sum_{m=0}^{n-1} p_m^{(n)} S^m, \quad \text{for} \quad n = 1, 2, \ldots N+1$$

For mathematical completeness, the condition $p_m^{(n)} \equiv 0$ for $m < 0$ or $m \geq n$.

Also, $P_0^{(1)} = 1$ since $P_1 = 1$.

The steps to be followed for calculation of coefficients are given below:

1. Define the initial state as
   $p_{m=-2\ldots N}^{n=0\ldots N+1} = 0$, n is the number of layers.
2. Apply $P_1 = 1$ setting $P_0^{(1)} = 1$.
3. Calculate the coefficients using the following expression:

$$p_m^{(n)} = Z_cY_{n-1}p_{m-1}^{(n-1)} + 2p_m^{(n-1)} + p_{m-2}^{(n-2)} - p_m^{(n-2)}$$

for $n = 2 \ldots \ldots N+1$, and $m = 0, 1, \ldots n-1$.

Calculation of Coefficients of $N_\Gamma(S)$

The coefficients a_n in the following expansion, i.e.,

$$N_\Gamma(S) = \sum_{n=1}^{N} a_nS^n - 1 = (Z_cS + 1)P_N - P_{N+1}$$

can be solved using

$$a_n = Z_cp_{n-1}^{(N)} + p_n^{(N)} - p_n^{(N+1)}, \quad \text{for} \quad n = 1, 2, \ldots N$$

Appendix D: Smith Chart-Based Calculations for a Jaumann Absorber

Computation of Ground Plane Admittance
Ground plane admittance, as seen at the plane of the resistive sheet looking toward the ground plane, can be calculated using the following steps:

Step 1: Calculate the ground plane admittance using

$$Y_g = \frac{-j}{Z_m} \cot\left(2\pi d\, \frac{\sqrt{\varepsilon_r}}{\lambda}\right)$$

Step 2: Plot the admittance values on the Smith chart.

Alternatively, if one has impedance values, then these impedance points on Smith chart can be transformed to admittance points by rotating 180° in clockwise direction, as shown in Fig. D.1.

When admittance points for all frequencies of desired frequency range are plotted, the admittance curve is obtained (Curve 1). Figure D.2 shows the

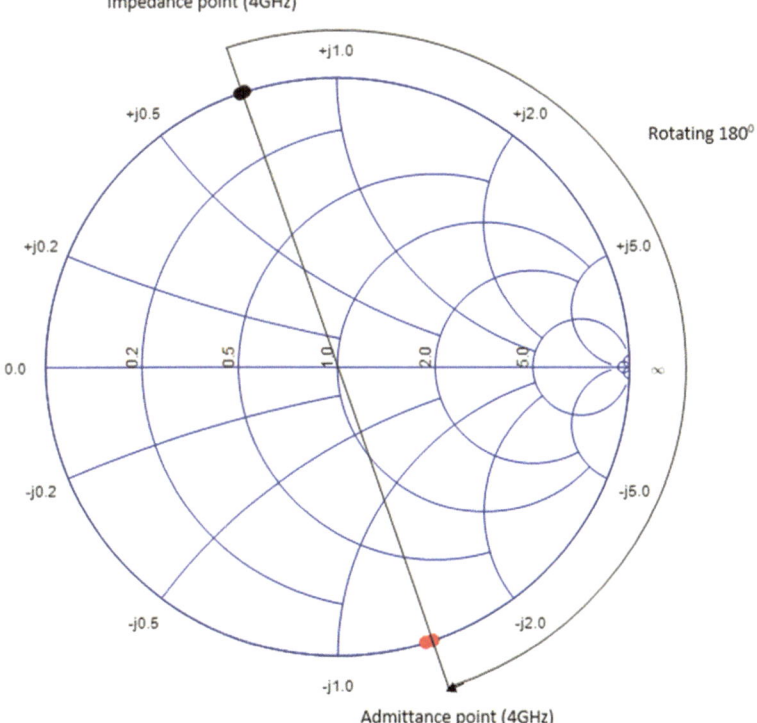

Fig. D.1 Admittance point on Smith chart (4 GHz)

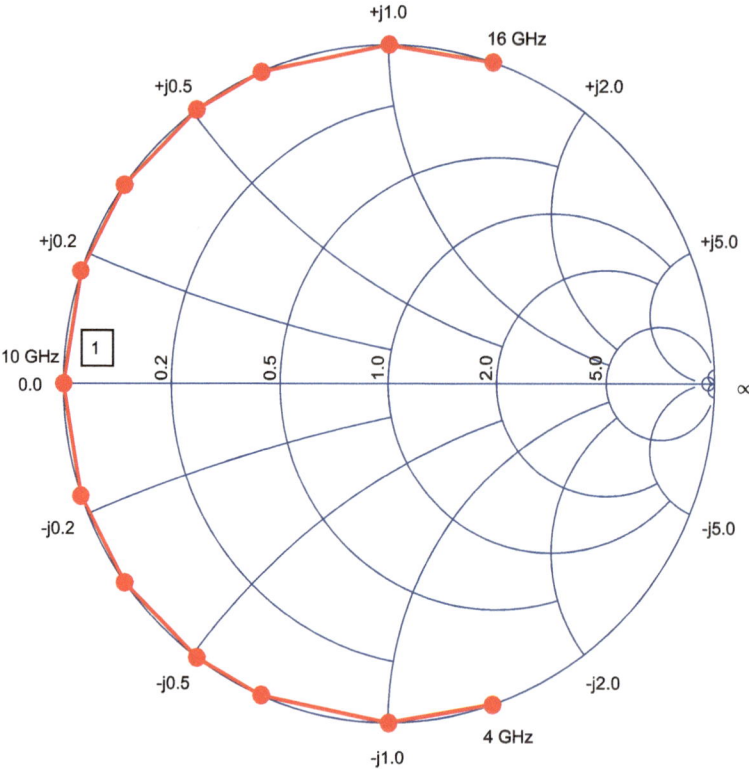

Fig. D.2 Admittance of the ground plane

admittance curve of ground plane for frequency range 4–16 GHz. It is to be noted that 10 GHz is the center frequency.

Computation of Resistive Sheet Admittance

Step 3: Plot the impedance of sheet on Smith chart. (Here, sheet impedance is $0.58Z_0$.)

Step 4: Obtain corresponding admittance value, by rotating the impedance value by 180° (Point 2). This is shown in Fig. D.3.

Computation of Admittance Above Resistive Sheet

Step 5: Add admittance of Curve 1 and Point 2 resulting in Curve 3, shown in Fig. D.4.

Computation of Admittance at Dielectric Layer Above the Resistive Sheet

The admittance at dielectric layer above resistive sheet is obtained by traveling through the dielectric layer of thickness d_2.

Step 6: Add the admittance of the second dielectric layer to the center of the Smith chart and mark that point as T.

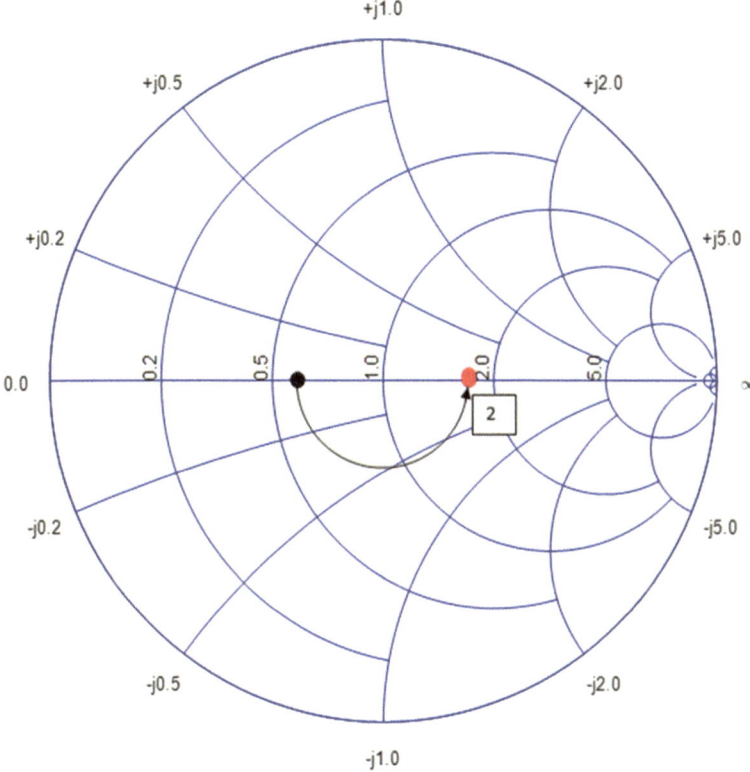

Fig. D.3 Resistive sheet admittance

Step 7: Define *path difference* as $\lambda_2/4$ where λ_2 is the wavelength corresponding to the center frequency in the second dielectric layer, i.e., $\lambda_2 = \lambda_0/\sqrt{\varepsilon_r}$.

Step 8: Compute the phase difference at every frequency using the *Phase Difference* $= \left(\frac{2\pi}{\lambda/2}\right)$ *Path Difference*

where λ is the wavelength of corresponding frequencies in the second medium. Since the Smith chart is designed for the analysis of a circuit at half-wavelength, the phase difference for half-wavelength is calculated.

Step 9: For a particular frequency, draw a line from T to the corresponding point on Curve 3. Rotate this line by the angle obtained in Step 8, taking T as center in the anticlockwise direction. Intersect the new line at a distance equal to the distance between T and the corresponding point of Curve 3.

Step 10: Repeat Step 9 for all the frequencies and obtain Curve 4.

It is to be noted that the rotation is more for higher frequencies than $\lambda_2/4$ (at center frequency) and less than $\lambda_2/4$ for lower frequencies. This is shown in Fig. D.5.

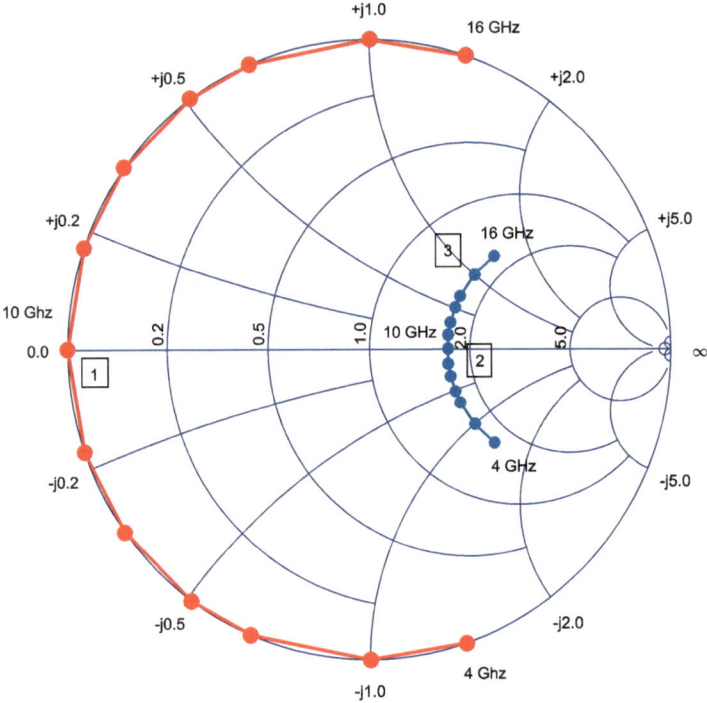

Fig. D.4 Admittance above resistive sheet (Curve 3)

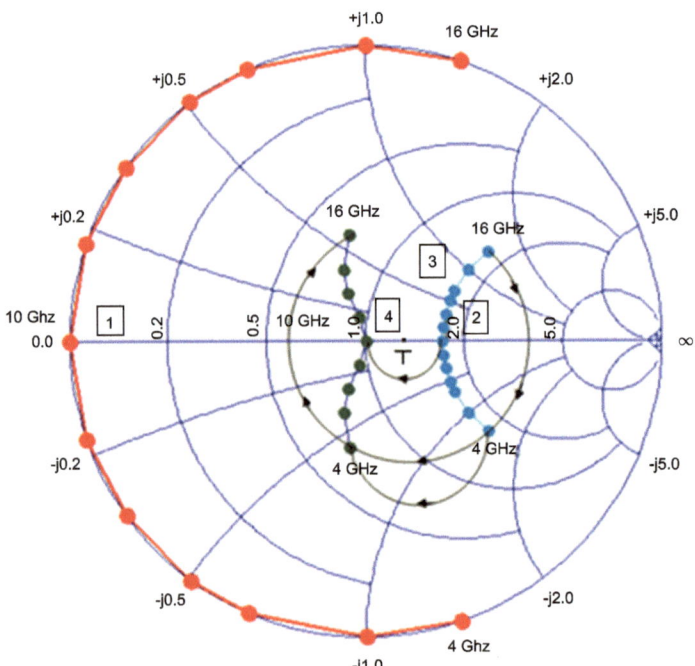

Fig. D.5 Rotation of admittance values of Curve 3

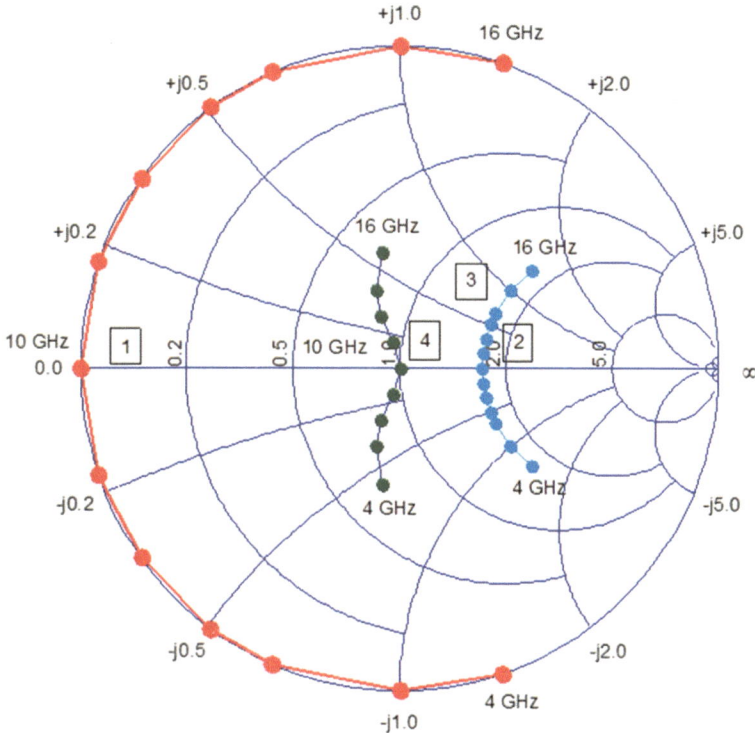

Fig. D.6 Admittance of a Jaumann absorber

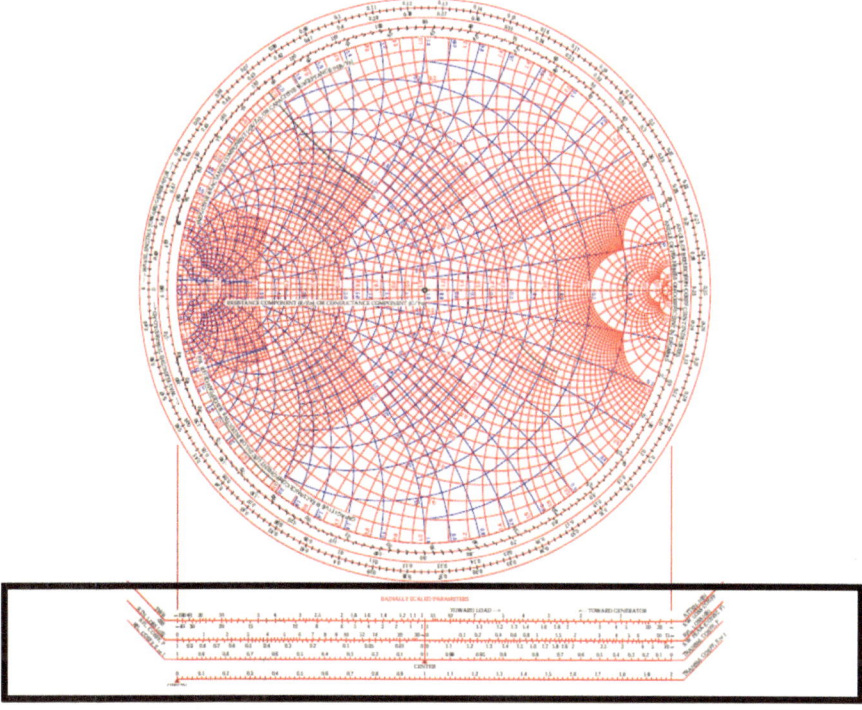

Fig. D.7 Smith chart

This rotation of admittance values gives Curve 4, as shown in Fig. D.6. It can be observed that frequencies are more clustered in Curve 4 than in Curve 3.

These admittances can then be used to obtain reflection coefficient of the absorber. This is available as linear scale in each Smith chart, as shown in Fig. D.7.

Authors' Biography

Dr. Hema Singh is currently working as principal scientist at the *Centre for Electromagnetics* of CSIR-National Aerospace Laboratories, Bangalore, India. She was lecturer in EEE, BITS, Pilani, India, in 2001–2004. She obtained her Ph.D. degree in electronics engineering from IIT-BHU, Varanasi, India, in 2000. Her active area of research is computational electromagnetics for aerospace applications, EM analysis of propagation in an indoor environment, phased arrays, conformal antennas, and radar cross section (RCS) studies including active RCS reduction. She received the Best Woman Scientist Award in CSIR-NAL, Bangalore, in 2007–2008 for her contribution in the area of active RCS reduction. Dr. Singh has authored or co-authored 11 books, 1 book chapter, 7 software copyrights, 235 scientific research papers and technical reports. She has also supervised over 40 graduate projects and postgraduate dissertations.

Mr. Ebison Duraisingh Daniel J obtained his B.Tech. (ECE) and M.Tech. (communication systems) from Karunya University, Coimbatore, India. He was a project scientist at the Centre for Electromagnetics (CEM) of CSIR-National Aerospace Laboratories, Bangalore, India, where he worked on RCS studies for aerospace vehicles.

Harish Singh Rawat is research scholar in Apex Level Standards & Industrial Metrology Department of CSIR-NPL, New Delhi. He has obtained his M.Sc. (Electronics) in 2013 from the Department of Engineering and Technology, Jamia Millia Islamia, New Delhi, India. He was a project engineer at the Centre for Electromagnetics (CEM) at CSIR-National Aerospace Laboratories, Bangalore, India. His research interests include radar cross section (RCS) studies, radar absorbing structures (RAS), low-profile antennas, and phased arrays mounted on planar and non-planar surfaces.

Ms. Reshma George obtained her M.Tech. in power electronics and drives in 2015 from the Department of Electrical and Electronics Engineering, Anna

© The Author(s) 2018
H. Singh et al., *Fundamentals of EM Design of Radar Absorbing Structures (RAS)*,
DOI 10.1007/978-981-10-5080-0

University Regional Centre, Coimbatore, India. She obtained her B.Tech. in electrical and electronics engineering in 2012 from MG University, Kottayam, Kerala. She is NALTech project scientist at the Centre for Electromagnetics (CEM) of CSIR-National Aerospace Laboratories, Bangalore, India, where she worked on radar absorbing structures (RAS) and conformal arrays.

Author Index

B
Bozzi, M., 1

C
Chen, M., 20
Costa, F., 19

E
Elia, U.F.D., 27

F
Fang, D., 20

K
Karlsson, A., 3
Kazemzadeh, A., 3
Ke, L., 28

L
Lei, H., 20

M
Manara, G., 19
Monorchio, A., 19
Munk, B.A., 1, 27
Munk, P., 3

P
Pelosi, G., 27
Peng, Z., 28

Perregrini, L., 1
Prior, J., 1, 27

R
Radmanesh, M.M., 27

S
Selleri, S., 27
Sjoiberg, D., 19
Sun, B., 29

T
Taddei, R., 27
Tang, Z., 29
Toit, L.J.D., 6, 7, 12, 15

W
Wang, Q., 29

X
Xinyu, H., 28
Xin, Z., 28
Xu, X., 29

Z
Zhang, Z., 20
Zhu, X., 20

Subject Index

A
Absorption, 1, 3, 11, 16, 21, 24, 30, 31, 37, 53
Admittance, 8, 10, 11, 14, 15, 19–21, 24, 29–34, 40–42, 44–50

B
Bandwidth, 1, 3, 11, 16, 30, 31, 37

C
Capacitance, 20, 23, 24

D
Dielectric layer, 4, 27, 46, 47

E
EM absorbers, 1, 3, 53
Equivalent circuit, 1, 2, 19–25, 37, 53

F
Frequency, 1, 6, 10, 16, 20–22, 24, 29–34, 37, 39, 45–47, 53
Frequency selective surface, 1

G
Ground plane, 1, 3, 20, 21, 25, 27, 29, 30, 33, 45, 46

I
Inductance, 20, 22–24
Input impedance, 6, 9, 31

J
Jaumann absorber, 1–11, 16, 17, 39, 40, 49

P
Perfect electric conductor, 4

R
Radar absorbing structure (RAS), 1, 3–17, 51, 53
Reflection coefficient, 5–7, 9–11, 13, 14, 16, 17, 22, 23, 25, 35, 50
Resistance, 20, 22, 23
Resistive sheet, 1, 3, 4, 10, 16, 20, 25, 27, 29–33, 45–48

S
Salisbury screen, 30–34, 45–48
Smith chart, 2, 27–35, 37, 45–47, 49, 50, 53
Spacers, 3, 10

T
Transmission line model, 1, 6, 20, 28

© The Author(s) 2018
H. Singh et al., *Fundamentals of EM Design of Radar Absorbing Structures (RAS)*,
DOI 10.1007/978-981-10-5080-0